Responses of Plants to Soil Flooding

Jun-Ichi Sakagami • Mikio Nakazono
Editors

Responses of Plants to Soil Flooding

Springer

Editors
Jun-Ichi Sakagami
Faculty of Agriculture
Kagoshima University
Kagoshima, Kagoshima, Japan

Mikio Nakazono
Graduate School of Bioagricultural Sciences
Nagoya University
Nagoya, Aichi, Japan

ISBN 978-981-99-9111-2 ISBN 978-981-99-9112-9 (eBook)
https://doi.org/10.1007/978-981-99-9112-9

This Springer imprint is published by the registered company Springer Nature Singapore Pte Ltd.
The registered company address is: 152 Beach Road, #21-01/04 Gateway East, Singapore 189721, Singapore

Preface

The growth of plants is greatly affected by both excesses and deficiencies in water. Environmental conservation, such as the greening of deserts, has previously received a large amount of research attention, including investigations related to drought stress. However, in comparison, relatively little attention has been paid to the study of plants experiencing excessive water stress. Temporary or permanent flooding is widespread in the world, and the importance of understanding this condition is increasing with the development of global warming. The two main types of flooding stress are waterlogging, in which roots are exposed to excessive water, and submergence, in which roots are flooded to the top of the shoots. Among the many physical and chemical changes, oxygen deprivation is the main factor inhibiting plant growth because of flooding. However, most previous studies on the responses of plants to flooding have focused on seed germination and soil water retention for the ease of experimental systems. In recent years, remarkable discoveries and developments have been recognized in the fields of molecular biology, plant physiology, and plant morphology in relation to submerged and waterlogged environments. The purpose of this book *Responses of Plants to Soil Flooding* is to introduce the latest research on flood-induced anaerobic responses to improve the availability and accessibility of this work for future research and general interest.

When rice plants are flooded, they grow taller and expand their leaves above the water surface. This is a survival strategy to escape the anaerobic environment and maintain aerobic energy metabolism and photosynthesis, by expanding their leaves above the water surface. In water, the solubility of gases such as oxygen and carbon dioxide are low and their diffusion rate is 10^4 times slower than in the atmosphere. Therefore, in flooded conditions where the water surface is stagnant, the oxygen concentration in the water becomes low due to the respiration of green algae, especially at night, and the hypoxia affects the energy metabolism of rice. Under such hypoxic conditions, rice undergoes anaerobic metabolism to obtain energy. However, anaerobic metabolism produces only two moles of ATP from one molecule of glucose, which is much less energy efficient than the 38 moles of ATP produced by aerobic metabolism. Therefore, if the anaerobic environment is prolonged, carbohydrates will be depleted, and the rice plant will not be able to survive. In addition, the supply of carbohydrates by photosynthesis in the water is also reduced due to the inhibition of light penetration and the decrease in carbon dioxide concentrations in the plant. By quickly escaping this flooding stress environment and expanding the leaves above the water surface, it is possible for the plant to take in oxygen and carbon dioxide from the atmosphere and improve the photosynthetic capacity of the leaf blade. The most important features of this work are to analyze the mechanisms of these types of physiological plant responses to anoxia and hypoxia under flooding conditions. Specifically, the importance of photosynthetic functions, chlorophyll injury, photosynthetic product translocation, root plasticity, and respiration in plants under anaerobic conditions.

In this book, leading researchers from this field plan to systematically explain the latest research results on anaerobic responses in plants. The book is characterized by its in-depth coverage of the tolerance functions of plants under hypoxic conditions from a variety of perspectives, with a particular focus on research areas related to molecular biology, physiology, and genetics. Although drought responses have dominated water stress research in the past, recent floods and attempts to introduce new cropping systems have made it necessary to take measures against excessive water stress injury, and systematic research is needed for this purpose. From this point of view, the approach taken in this book is new and interesting in that it covers basic research and adaptation technologies in the field and can be applied to various different situations. Avoiding the effects of global warming, which are expected to increase in the future, and improving the adaptability of plants to the environment are two extremely important and current issues. The focus of this book is how plants can adapt to poor environments and improve productivity under the conditions of soil hypoxia caused by excess water, such as heavy rains and typhoons. From this point of view, the reader will be able to understand the various adaptations of plants to climate change, which will clarify the future directions of research and show the possibility of applying the knowledge and techniques gained in this book to the field.

The Sixth Report of the Intergovernmental Panel on Climate Change (2021) warns that global warming will proceed faster than previously assumed and that all regions of the world will face increasing changes. "Climate Resilient Development" has been proposed as a key phrase to combat global warming, and it is important to identify the adaptive capacity of plants and improve it where possible. In this regard, the publication of this book, which includes ideas for mitigating flood damage caused by global warming, is extremely important, timely, and rational. A portion of our book may be suitable for undergraduate courses, but our audience is primarily scholars. Writing on the narrow topic of plant anaerobiosis for a well-defined audience will help our book resonate more strongly with its readers and increase the impact of this book.

Kagoshima, Japan Jun-Ichi Sakagami
Nagoya, Japan Mikio Nakazono

Contents

Part I
Impacts of Climate Changes on Plant Adaptation

Chapter 1
Abiotic Factors Affect Plant Growth

Tohru Kobata

1.1 Increased Impacts of Abiotic Factors Such as Flooding, Drought, High Temperature and Soil Deterioration on Crop Production Under Climate Change

Crop plants suffer from flooding, drought, high temperature, and soil degradation, such as soil compaction, salinity, acidification, and metal pollution, and these abiotic factors seriously inhibit crop production (Loomis and Connor 1992). Climate change has modified the total rainfall and distribution of precipitation in seasonal rainfall, resulting in water availability extremes, including drought and flooding in agricultural regions of the world in the past 50 years (Bailey-Serres et al. 2012; Shortridge 2019; Kaur et al. 2020). The model estimations show that areas of hyper-arid, arid, semiarid, and/or dry subhumid are expanding with generally greater aridity at a global scale, but at regional and local scales, the outcomes may vary, with dryland areas decreasing in some temperate regions and expanding in tropical and subtropical regions (Hermans and McLeman 2021). Furthermore, models suggest a worst world scenario in which small floods for farm water supplies are decreasing but large flood events that pose a risk to life and infrastructure are increasing (Wasko et al. 2021). Therefore, the instability of rainfall as one of the critical abiotic factors for crop production has increased, and soil water conditions could become unfavorable to maintain crop growth in the future. As one of the most typical phenomena of climate change, the ranges of elevated temperature in the future differ depending on the scenario; the average global temperatures in 2100 may be 1.7–4.8 °C warmer than those in 2000 (Hermans and McLeman 2021). These changes in rainfall and temperature could also have an impact on soil fertility and the physical and

T. Kobata (✉)
Faculty of Life and Environmental Science, Shimane University, Matsue, Japan
e-mail: kobata@life.shimane-u.ac.jp

J.-I. Sakagami, M. Nakazono (eds.), *Responses of Plants to Soil Flooding*,
https://doi.org/10.1007/978-981-99-9112-9_1

chemical conditions of soils: heavy rain reaches soil nutrients, a decrease in rainfall under arid and semiarid areas accelerates soil salinity by reducing salt reaching and upward groundwater containing salt under uncontrolled irrigation, and high temperature increases soil nutrient consumption by stimulating plant growth and decomposition of soil organic matter (Loomis and Connor 1992; Hiell 2003; Kaur et al. 2020). Therefore, the negative effects of abiotic factors on crop production could compound and become complicated under climate change.

1.2 Water Balance Causing Flooding and Drought in Crop Fields

Water is one of the most important abiotic factors. Flooding and drought that give serious impacts to crop production occur from an imbalance between supplied water and the output of water to soils in crop fields. The supplied water consists of rainfall (W_R), irrigation (W_I), and upward water supply from the groundwater table (W_G) (Fig. 1.1). Part of the supplied water ($W_R + W_I$) seeps or overflows (W_S), and the remaining water ($W_R + W_I - W_S + W_G$) gradually soaks into the soil sections at different depths. The soil section is a bulk soil with roots. Soil sections are regarded as a pool of water (W_{POOL}), which consists of plant available water (sometimes, it is defined as a fraction of transpirable soil water, FTSW) (Ray and Sinclair 1998) plus unavailable water fixed in soils (water content lower than wilting point) (Hiell 2003). The amount of W_{POOL} changes depending on the soil water-holding capacity determined by the soil physical properties and organic matter content (Hiell 2003). Water over the water-holding capacity is infiltrated out of the W_{POOL} (W_{IF}). Water in the soil sections is absorbed by plant roots as transpiration (T) and evaporated from

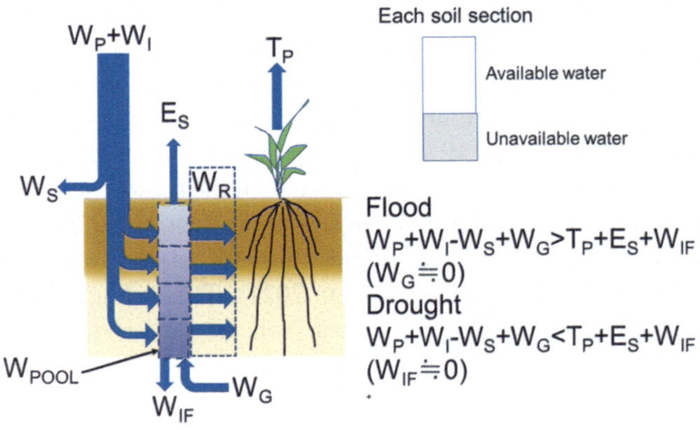

Fig. 1.1 Water balance of crop fields. W_P precipitation, W_I irrigation, W_S seeping or overflow, E_S soil evaporation, T transpiration, W_R root absorption, W_{IF} infiltration, W_G supply from ground water, W_{POOL} water pool of soil sections. Each soil section consists of available and unavailable water for plants. Cases of water balance during flooding and drought are shown

soil surfaces (T_S). Water absorbed by plant roots (W_R) equals transpiration (T) when absorbed, and transpired water is maintained in a steady state, while W_R temporarily differ from T (under high T, $T > W_R$ and under low T, $T < W_R$ by the buffer of water stored in the plant body).

Soil saturation and flooded soils occur when the water supply to crop fields ($W_P + W_I - W_S + W_G$) is over the maximum potential of W_{POOL}, as the water use of the fields ($T_P + E_S - W_{IF}$) is less than W_{POOL}. When $W_P + W_I$ continuously increases and W_S reduces, air spaces in soils are filled by water, and furthermore, crop plants are submerged. Meanwhile, drought is the opposite case in which $W_P + W_I - W_S + W_G$ is below $T + E_S - W_{IF}$. W_G under flooding and W_{IF} under drought can be ignored because upper water flow under flooding conditions would be scarce, and under drought conditions, water movements would be small due to the high resistance of water flux in soils unless a high water table continuously supplies water (Hiell 2003).

Climate change could alter these water economies due to an increase or decrease in the amounts of rainfall (W_P) and evapotranspiration ($T + E_S$) related to radiation and humidity (Loomis and Connor 1992) and W_{POOL} related to soil physical properties influenced by the contents of organic matter (Hiell 2003).

1.3 Determination Factors for Crop Yield

Crop yield (Y) is indicated by

$$Y = \mathrm{BY} \times \mathrm{HI} \tag{1.1}$$

where BY is the biomass yield and HI is the harvest index (Passioura 1977; Unkovich et al. 2010). From the equation, yield is determined by two factors, BY and HI. In crops with harvested biomass such as sugarcane, BY nearly equals Y. The effect of abiotic factors such as flooding, drought, high temperature, and soil degradation on Y can be reviewed from the effects on BY and HI.

1.4 Effects of a Single Abiotic Factor on Crop Production

1.4.1 Flooding

Flooding causes water saturation of soils and submergence of plants; water saturation of soils removes oxygen in soil pores and directly inhibits root respiration; and furthermore, submerged water prohibits leaf gas exchange, decreasing plant growth and causing plant death, while the timing and length of flooding change plant damage (Bailey-Serres et al. 2012; Kaur et al. 2020). The negative effects of water submergence and saturation of soil pores on gas exchanges such as photosynthesis and respiration depend on a much lower diffusion rate of CO_2 and O_2 in water than in air, in which these gases are 1/10,000th and the flux of O_2 into soils is approximately 320,000 times less in water-filled soil pores than in gas-filled soil pores (Kaur et al. 2020).

The effect of flooding on yields in wheat and soybean under variable terms at different growth stages is summarized by Kaur et al. (2020), and these data are shown in figures (Fig. 1.2). Yield loss (flood/control) in wheat tends to increase around the flowering stage in the reproductive stage, although in the vegetative stage, the extension of flooding terms increases the effect. In soybean, of which the reproductive stage (R1-) starts from the early growth season, yield loss occurs across the whole growing season, while fluctuations in the loss are observed. In rice plants that can grow under flooded conditions, clear yield loss is not observed except in cases of heavy submergence of aboveground plants (Suge 1987; Sakagami et al. 2012). The high adaptation of rice plants to flooded soils comes from the development of aerenchyma in roots that supplies O_2 to root systems (Arikado 1955). Aerenchyma development is one type of the flood resistance methods often observed in crop plants (Bailey-Serres et al. 2012). Aerenchyma tissue composed of low-resistance gas conduits in roots and stems enables diffusion and exchange of oxygen and carbon dioxide from near the root apex to the uppermost submerged region of the root and into the stem (Jackson and Armstrong 1999). Adventitious roots that emerge from stem tissue under conditions of partial to complete submergence can replace compromised roots and provide efficient aerenchymatous connections between aerial shoot tissues and submerged organs (Bailey-Serres et al. 2012). In upland crops such as wheat and soybean, the development of aerenchyma after flooding has been observed, although it could be insufficient for growth maintenance (Arashi and Nitta 1955; Arikado 1955; Bailey-Serres et al. 2012). Plants can

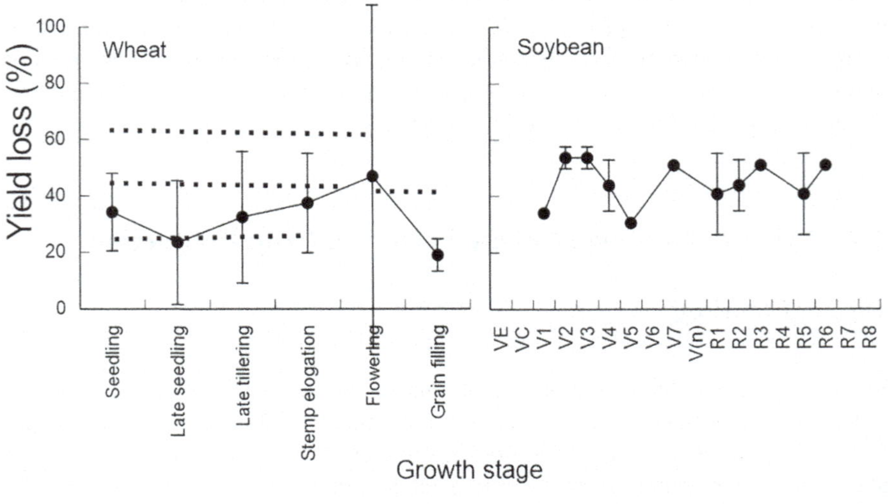

Fig. 1.2 The effect of flooding on yield loss in different growth stages in wheat and soybean. Data are the average of the loss under variable terms of flooding treatments (1–60 days in wheat and 1–14 days in soybean) and vertical bars are standard deviations of observed data derived from Kaul et al. (2020)

escape the submergence of the whole plant by stem elongation, such as deep water and flooded rice (Bailey-Serres et al. 2012), and local rice cultivars in lowland areas along a river in Japan which often suffer from flooding have a high capacity for stem elongation (Suge 1987).

Under flooding conditions, leaf area, branching, plant height, and carbohydrate contents generally decrease in diverse crops (Kaur et al. 2020), and hence, these effects of flooding could have a strong impact on BY. In rice, submergence and flooding reduce plant height and tiller number related to BY but have less influence after heading (Sakagami et al. 2012). The higher yield loss around flowering in wheat suggests that the HI reduces Y due to the inhibition of fertilization (Olgun et al. 2008). These results suggest that flooding mainly reduces Y due to the effect on BY, although a low HI due to low fertilization and fewer flower organs influences Y.

An increase in W_S and W_{IF} under flooding increases the reaching of nitrogen (Kaur et al. 2020); approximately 10–40% of applied N can be lost through leaching depending upon the application time, fertilizer source, and the crop being fertilized (Legg and Meisinger 1982; Kaur et al. 2020). Furthermore, ammonia (NH_3) volatilization increases due to changes in soil microbial communities under flooding (RafaelSánchez-Rodríguez et al. 2019). These nutrient losses immediately lower BP related to Y loss. Flooding effects on crop yields can be lightened by cultivation manipulations such as draining and raised beds (high ridges) to increase the W_{IF} and nitrogen application (Kaur et al. 2020).

1.4.2 Drought

Drought has been a dominant limiting factor for crop production not only in arid and semiarid areas (Turner 1996) but also in humid zones where rainfall has a seasonal bias (Tazaki et al. 1980). The physiological mechanisms, genotypic differences, and genetic background of drought in crop plants have been precisely analyzed and reviewed (Hsiao 1973; Begg and Turner 1976; Turner 1986; Dietz et al. 2021; Panda et al. 2021). The drought resistance of crop plants is conceptually divided into drought tolerance, avoidance, and escape (Begg and Turner 1976); these three traits contribute to maintaining BY and/or assimilating the location into harvest plant organs based on the HI.

Water use can be related to BY by

$$\mathrm{WUE} = \mathrm{BY} / \left(T / \mathrm{VD}\right)$$

where WUE is the water use efficiency and VD is the vapor water deficit of air (Sinclair et al. 1984). From the equation,

$$\mathrm{BY} = \mathrm{WUE} \times \left(T / \mathrm{VD}\right) \tag{1.2}$$

This equation indicates that higher WUE and/or T/VD can maintain BY. Higher WUE is an effective trait for increasing BY under limited water supply. However, improvements in transpiration efficiency of a single leaf (photosynthesis rate/transpiration rate), as the water use efficiency of a leaf-level base correlates well with carbon discrimination, do not always translate into higher WUE of the whole plant base or yield in wheat (Condon et al. 2004). Furthermore, high shoot dry matter increases in rice cultivars under terminated irrigation highly depended not on the superiority of WUE but on the plant water use (T), and the higher T was supported by higher root distribution into moist soil layers (Kobata et al. 1996; Kobata and Palta 2018). In fact, the variation in WUE among genotypes was not larger than that in T/VD. Therefore, in crop plants, there is controversy over whether WUE can contribute to the maintenance of BY (Blum 2005). T/VD is high, if T is higher and/or VD is lower. A lower VD is realized under high humidity conditions and low absolute humidity under cool temperatures. Hence, the planting season is important for escaping from high VD, while the terminal crop season is sometimes under high-temperature and low-humidity conditions, such as wheat in the Mediterranean zone and summer crops (Tazaki et al. 1980: Sadras and McDonald 2012). Higher T results from maintaining leaf area and leaf stomatal conductance under drought; maintenance of leaf expansion, green leaf area, and stomatal apparatuses under desiccated soils is needed. Escape from plant dehydration under drought is one of the most effective traits for maintaining these plant growth processes related with higher T (drought avoidance) (Begg and Turner 1976; Blum 2005), although osmotic adjustment as a capacity of plant adaptation can maintain these processes of dehydrated plants (Turner 1996; Turner et al. 2014), and partial soil desiccation around roots makes it possible to close stomata by hormonal signals even in nondehydrated leaves (Turner 1996). High root accesses of roots with extensive W_{POOL} can increase T. A large difference in root distribution in deep soil layers was observed among cultivars in rice (O'Toole and Bland 1987) and wheat (Mian et al. 1994); the distribution was one of the most important key factors for maintaining T to escape from plant water deficit in crops (Turner 1996; Kobata et al. 1996; Kobata and Palta 2018; Kobata et al. 2018b).

How does soil desiccation under drought conditions relate to T? The relative T (T/T_0, where T_0 is transpiration in the well-irrigated control) is well fitted by the equation for FTSW (Ray and Sinclair 1998).

$$T / T_0 = f(\text{FTSW}) \tag{1.3}$$

$$\text{FTSW} = (\text{SW} - \text{SW}_\text{W}) / (\text{SW}_\text{F} - \text{SW}_\text{W})$$

where SW is the observed soil water content, SW_F is the upper limit of water absorption (field capacity), and SW_W is the lower limitation (wilting point), while SW_F and SW_W are not physically defined but empirical (Hiell 2003). Eq. (1.3) was not influenced by root size in soybean and maize (Ray and Sinclair 1998). A case in several rice plants for the relationship between T/T_0 and FTSW has been shown (Kobata, unpublished data) (Fig. 1.2). In the relationship, T/T_0 started to decrease at similar

FTSWs in diverse cultivars and different soil types (volcanic and sandy soil). From Eqs. (1.2) and (1.3), the higher FTSW contributes to the maintenance of T related to BY. The higher FTSW can be maintained by a deep root system distributed into large wet bulk soils (larger W_{POOL}).

Suppression of E_S saves water in W_{POOL} and can reserve soil water for T. Covering the soil surface by the rapid expansion of leaves suppresses E_S (Kobata et al. 1996, 1998). However, the trait for the reduction of E_S and high WUE seems to not be compatible (Condon et al. 2004).

Drought at pollen development and the flowering stage strongly decreases fertility in many crops (Nagato 1949; Kobata et al. 1994; Sadras and McDonald 2012), and sterility profoundly decreases the HI by reducing grain number. Furthermore, a lack of assimilates during the grain-filling period under water shortage conditions decreases the HI due to inadequate grain filling (Kobata et al. 2018b). In crops suffering from terminal season drought, such as a typical case in the Mediterranean zone (Turner et al. 2014; Kobata et al. 2018b), the effect of a decrease in the HI on Y would become stronger (Figs. 1.4 and 1.5).

1.4.3 High Temperature

Temperature is a main factor that determines plant growth and development rates; phenological development is indicated by the developmental index (DVI) calculated by summing the developmental rate (DVR), in which DVR is a function of daily mean temperature with a day length (Horie et al. 1995). Elevated temperature shortens the vegetative and reproductive stages, resulting in inadequate BY and a small sink size of the yield portion, and furthermore, extreme high temperature injures plant organs such as leaves and flowers. The most serious effect of high temperature on Y is damage to pollens around the flowering stage, resulting in reductions in the HI; sterility occurs at temperatures over 35 °C in rice (Matsui and Hasegawa 2019) and soybean (Djanaguiraman et al. 2019) and 32 °C in wheat (Nuttall et al. 2018).

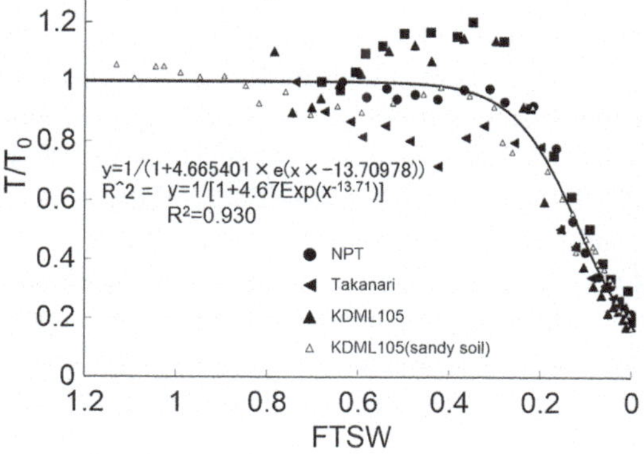

Fig. 1.3 Relative transpiration (T/T_0) and fraction of transpirable soil water (FTSW) in different rice cultivars when watering was terminated. T_0 is the transpiration of the well-watered control. Plants were grown in pots containing volcanic soil (Andosol), and one cultivar did in pots contained sandy soil (Unpublished data)

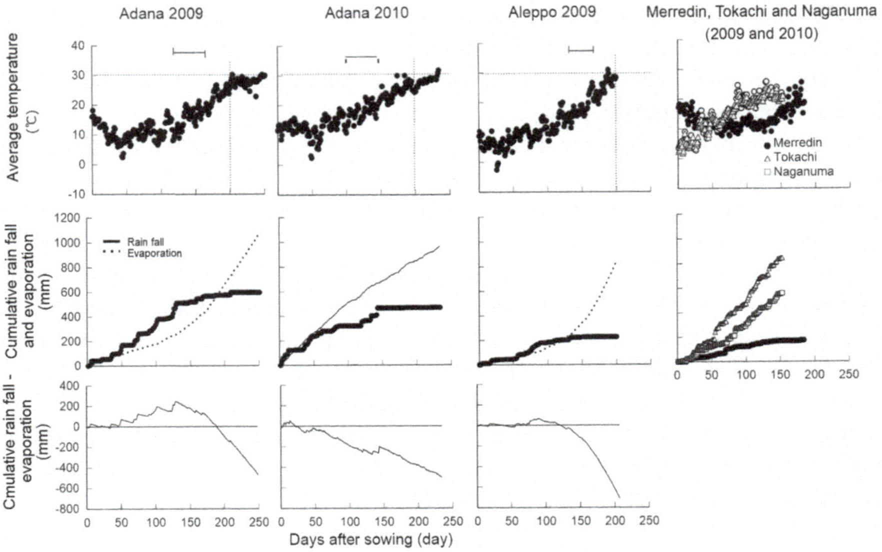

Fig. 1.4 Average temperature, cumulative rainfall, pan evaporation, and cumulative rainfall minus pan evaporation of the day after sowing in Adana, Turkey (2009 and 2010), and Aleppo, Syria (2010), in the wheat-growing season (Kobata et al. 2018b). Additionally, the average temperature and cumulative rainfall in 2009 and 2010 in Merredin, Western Australia, as a case of low temperature and rainfall, and in Obihiro and Naganuma, Hokkaido, Japan, as a case of slightly lower temperature to and high rainfall in wheat-growing areas, are shown

Hence, high-temperature tolerance for fertility or escape from high temperature at flowering would be important traits for high-temperature resistance (Ishimaru et al. 2022). The high-temperature resistance across the whole growth period would be even more important for crop yield because of recent temperature elevation events during the whole growth season (Hermans and McLeman 2021).

1.4.4 Other Abiotic Factors

The elevated CO_2 that is one of the causes of climate change increased BY and *Y* in 186 independent studies of 18 C_3 crops based on free-air CO_2 enrichment (FACE) experiments (Ainsworth and Long 2021), although acclimation of plants was possible to reduce the effect of elevation on BY in the long term (Tausz-Posch et al. 2020). Under elevated CO_2, Eq. (1.3) did not clearly differ from that under ambient conditions in spring wheat (Kobata 2006).

Other abiotic factors, such as soil compaction and nutrient, chemical, biological, and physical properties that impact crop production, are affected not only by regional soil characteristics but also by climate conditions such as rainfall and by cultivation procedures such as tillering and irrigation (Loomis and Connor 1992;

Hiell 2003). Therefore, under climate change, the effect of these factors on crop plants could be integrated and intensified.

1.5 Combined Effects of Abiotic Factors on Crop Production

Crops suffer from multiple abiotic stresses under climate change (Hermans and McLeman 2021). Temperature is a widespread factor that acts simultaneously with other abiotic factors; hence, the combined effects of high temperature with flooding and drought on crop production are addressed.

High temperature accelerates plant damage under flooding as a decrease in plant survival in corn plants (Nielsen 2015) and grain yield in winter wheat (Wu et al. 2014) and soybean (Wuebker et al. 2001). These negative effects of temperature elevation result from a reduction in soil oxygen due to an increase in the respiration of microorganisms and plants (Kaur et al. 2020). Furthermore, under higher temperature conditions, flooding induced a rapid release of nutrients and resulted in changes of soil microbial biomass and microbial community structure to influence soil fertility and physical properties related to water holding (RafaelSánchez-Rodríguez et al. 2019; Tisdall 1994). Therefore, high temperature intensifies the negative impacts on crop productivity under flooded conditions. Fertilizer manipulations, such as nitrogen application, cultivated operations such as draining and high-ridge cultivation, and genetic improvements, would be highly required to lessen flooding damage under climate change.

The integrated effect of high temperature and drought on wheat production is suggested in a case in the Mediterranean zone in which rainfall terminated under high-temperature conditions. In southwestern Turkey (Adana) and northern Syria (Aleppo), low rainfall, high evaporation, and high temperature start in the middle growing season, and wheat suffered from severe terminal drought and high temperature (Fig. 1.5) (Kobata et al. 2018b). High temperature increased evaporation due to an increase in VD and accelerated the soil water deficit. The severe high temperature and low rainfall in these two areas were proven to be typical from a comparison with other main wheat-growing areas, Merredin in the Mediterranean zone in western Australia (Bureau of Meteorology 2022), where rainfall was similarly low but temperature was lower, and Obihiro and Naganuma in the subarctic zone in Hokkaido, Japan (JMA 2022), where temperature was slightly lower but rainfall was much higher. In Adana and Aleppo, three different types of wheat, bread wheat (*Triticum aestivum* L.), durum wheat (*T. durum*), and emmer wheat (*T. dicoccum*), which were selected according to their localities, origins, and genetic diversity, were grown under rainfed conditions. The grain yield was closely correlated with the HI across different types of wheat in each location but was not correlated with BY (Fig. 1.5), and hence, *Y* was significantly determined by the partitioning of assimilates into the grain. The *Y* in bread wheat had a higher HI than the others. The HI was negatively correlated with heading days after sowing (Fig. 1.6). Delay of day of heading after sowing increased both average temperature

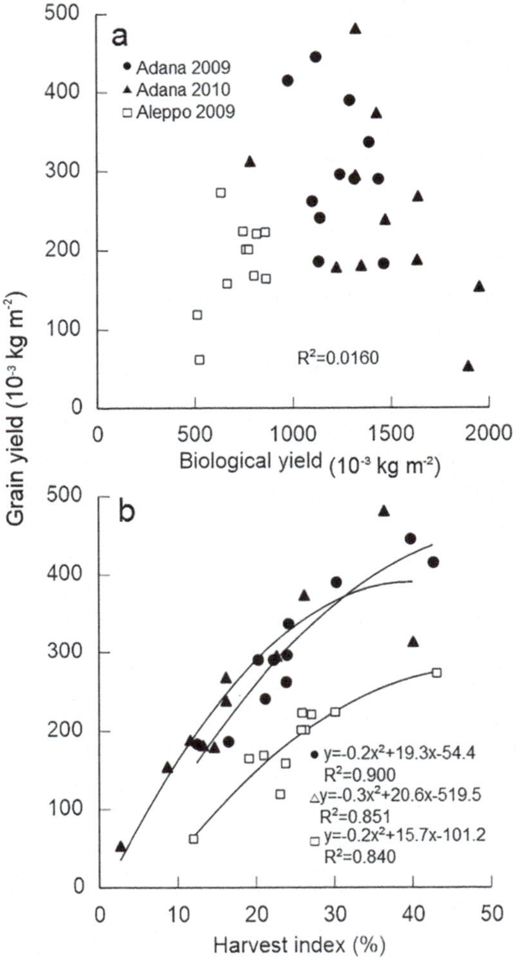

Fig. 1.5 Relationships between grain yield and biomass production (a) and harvest index (b) in three wheat types grown in Adana and Aleppo in Fig. 1.4 (Kobata et al. 2018b)

during the grain-filling period (0–30 days after heading) and drought strength indicated by amounts of rainfall minus evaporation. Cultivars with shorter days between sowing and heading, such as bread wheat, have the benefit of escaping terminal drought and high temperatures. The higher contribution of the HI to *Y* would come from the impact of high temperature and drought at the terminal stage on fertilization around the flowering stage, resulting in sterility (Nuttall et al. 2018; Fábián et al. 2019) and inadequate grain filling; the high temperature accelerates the grain-filling rate under limited assimilation under drought, and as a result, grain filling decreases and finishes inadequately (Egli 2004; Kobata et al. 2012, 2018a). In Adana, a growth model estimated that precipitation will be further reduced in the future and wheat production will become serious (Nakagawa et al. 2006).

In Merredin with lower temperature and low rainfall and Naganuma with similar temperature and higher rainfall, *Y*, BY, and HI were 304 gm^{-2}, 750 gm^{-2}, and 41

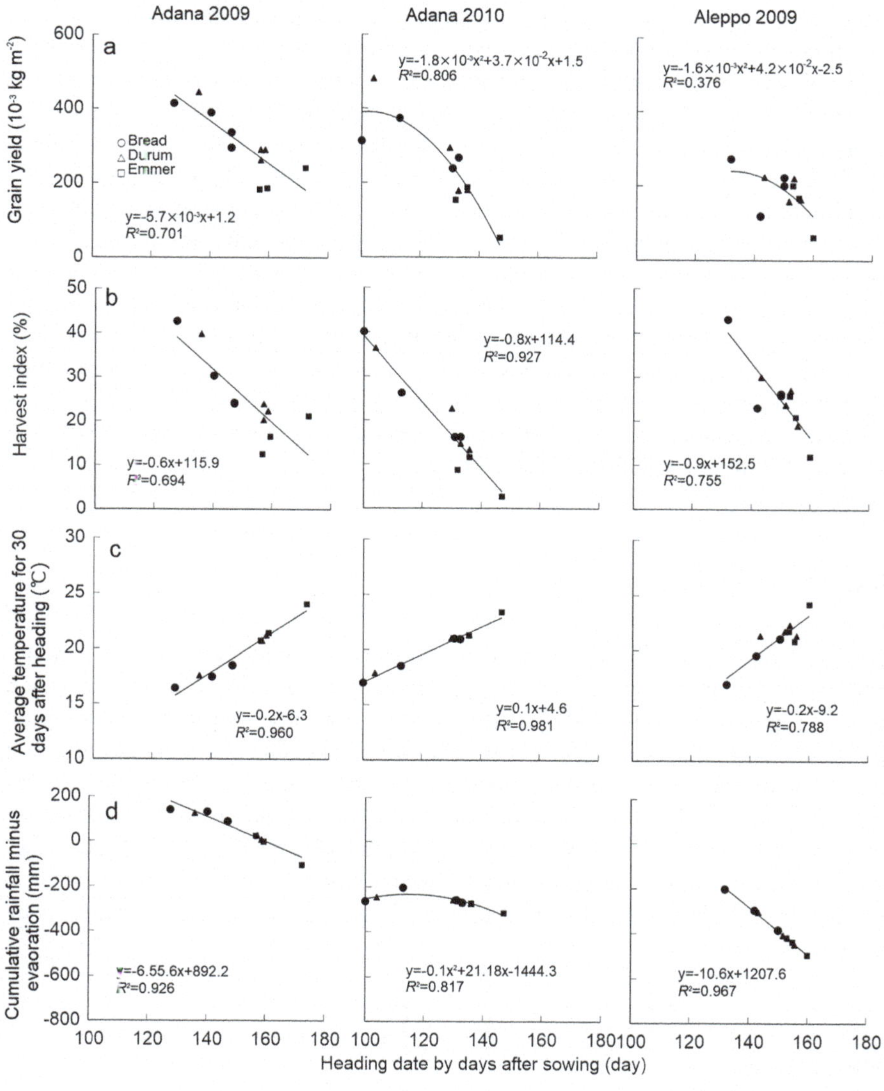

Fig. 1.6 The effect of heading date by days after sowing on grain yield (**a**), harvest index (**b**), average temperature for 30 days after heading (**c**), and rainfall minus evaporation (**d**) in three types of wheat in Adana and Aleppo in Figs. 1.4 and 1.5 (Kobata et al. 2018b)

(for 9 cultivars at 23 experiments in 2003–2007) (Sharma et al. 2008) and 426, 1027, and 38, respectively (for two cultivars in 2003–2006) (Hossain et al. 2010). From these results, the relationship between Y and HI in low-temperature and high-rainfall areas was approximately similar to that in high-temperature and low-rainfall zones such as Adana and Aleppo, while high BY in Naganuma under high rainfall did not result in high Y due to a lower HI. Therefore, assimilate partitioning into grains indicated by the HI would be a key factor for Y under diverse temperature and water supply conditions in wheat. Similarly, it was suggested that in modern high-yielding cultivars of soybean for rainfed semiarid environments, higher HI has been selected with lower biomass and leaf area and larger grain size (Jin et al. 2019).

Simultaneous elevation of CO_2 and temperature in the Chinese double rice-cropping system (Wang et al. 2020) and elevation of CO_2 with drought in wheat (Ulfat et al. 2021) increased BY, while the elevation of CO_2 seems to intensify requirements of soil nutrition (Tausz-Posch et al. 2020). Other abiotic factors, such as soil nutrient and chemical conditions, biological components, and physical properties, could be influenced by rainfall and temperature under climate change and indirectly integrate long-term effects on crop production (Marschner 1995). Hence, the responses of crop plants to these abiotic factors should be more carefully evaluated and assessed under conditions close to actual plant cultivation as much as possible, because the interaction between aboveground plants or roots and root sizes under experimental conditions sometimes causes critical effects on crop growth and water use (Poorter et al. 2012; Turner 2019).

1.6 Conclusion

Abiotic factors, typically represented by flooding, drought, elevated temperature, and soil degradation, have serious impacts on crop production, while the elevation of CO_2 is possible to have positive effects, and the negative impacts could increase by the integration of abiotic factors with future climate change. It should be noted that flooding increases global warming because of an increase in greenhouse gas emissions (RafaelSánchez-Rodríguez et al. 2019). Abiotic stress decreases crop yields due to suppression of BY in many cases, and in some cases, the HI decreases due to sterility and poor grain filling. Furthermore, long-term effects on crop production by abiotic factors, as represented by soil degradation under climate change, are considered.

References

Ainsworth EA, Long SP (2021) 30 years of free-air carbon dioxide enrichment (FACE): what have we learned about future crop productivity and its potential for adaptation? Global Change Bio 27:27–49

Arashi K, Nitta H (1955) Studies on the lysigenous intercellular space as the ventilating system in the culm of rice and some other graminaceous plants. Jpn J Crop Sci. 24:78–81

Arikado H (1955) Studies on the development of the ventilating system in relation to the tolerance against excess-moisture injury in various crop plants.: iv. the ventilating pressure in the rice plant growing in paddy-field conditions. Jpn J Crop Sci 23:285–286

Bailey-Serres J, Lee SC, Brinton E (2012) Waterproofing crops: effective flooding survival strategies. Plant Physiol 160:1698–1709

Begg JR, Turner NC (1976) Crop water deficits. Adv Agron 28:161–217

Blum A (2005) Drought resistance, water-use efficiency, and yield potential—are they compatible, dissonant, or mutually exclusive? Aust J Agric Res 56:1159–1168

Bureau of Meteorology (2022) Commonwealth of Australia. http://www.bom.gov.au/inside/index.shtml?ref=hdr

Condon AG, Richards RA, Rebetzke GJ, Farquhar GD (2004) Breeding for high water-use efficiency. J Exp Bot 55:2447–2460

Dietz KJ, Zörb C, Geilfus CM (2021) Drought and crop yield. Plant Biol 23:881–893

Djanaguiraman M, Schapaugh W, Fritschi F, Nguyen H, Prasad PV (2019) Reproductive success of soybean (*Glycine max L.* Merril) cultivars and exotic lines under high daytime temperature. Plant Cell Environ 42:321–336

Egli DB (2004) Seed-fill duration and yield of grain crops. Adv Agron 83:243–279

Fábián A, Sáfrán E, Szabó-Eitel G, Barnabás B, Jäger K (2019) Stigma functionality and fertility are reduced by heat and drought co-stress in wheat. Front Plant Sci 10:244

Hermans K, McLeman R (2021) Climate change, drought, land degradation and migration: exploring the linkages. Curr Opin Environ Sustain 50:236–244

Hiell D (2003) Introduction to environmental soil physics, 1st edn. Academic Press

Horie T, Nakagawa H, Centeno HGS, Kropff MJ (1995) The rice crop simulation model 4 SIMRIW and its testing. In: Matthews RB, Kropff MJ, Bachelet D, van Laar HH (eds) Modeling the impact of climate change on rice production in Asia. Wageningen University, pp 95–139

Hossain MA, Takahashi T, Jinno H, Senju K, Kawata Y, Zhang L, Araki H (2010) Grain filling mechanisms in two wheat cultivars, Haruyutaka and Daichinominori, grown in Western Japan and in Hokkaido. Plant Prod. Sci. 13:156–163

Hsiao TC (1973) Plant responses to water stress. Ann Rev Plant Physiol 24:519–570

Ishimaru T, Sasaki K, Lumanglas PD, Sasaki K, Lumanglas PD, Cabral CLU, Ye C, Yoshimoto M (2022) Effect of drought stress on flowering characteristics in rice (*Oryza sativa L.*): a study using genotypes contrasting in drought tolerance and flower opening time. Plant Prod Sci 25:359

Jackson MB, Armstrong W (1999) Formation of Aerenchyma and the processes of plant ventilation in relation to soil flooding and submergence. Plant Biol 1:274–287

Jin Y, He J, Turner NC, Du Y, Li H (2019) Water-conserving and biomass-allocation traits are associated with higher yields in modern cultivars compared to landraces of soybean [*Glycine max (L.)* Merr.] in rainfed water-limited environments. Environ Exp Bot 168:103883

JMA (2022) Japan meteorological Agency. https://www.jma.go.jp/jma/index.html

Kaur G, Singh G, Motavalli PP, Nelson KA, Orlowski JM, Golden BR (2020) Impacts and management strategies for crop production in waterlogged or flooded soils: a review. Agron J 112:1475–1501

Kobata T (2006) The effect of desiccated soils on transpiration in spring wheat subject to elevated temperature and CO_2 concentration. ICAAP Project. https://www.chikyu.ac.jp/iccap/ICCAP_Final_Report/5/2-crop_kobata.pdf

Kobata T, Palta JA (2018) An experimental irrigation tool for creating a water gradient across soil depths under terminated rainfall conditions. Irrig Sci 36:281–288

Kobata T, Tanaka S, Utumi M, Hara S, Imaki T (1994) Sterility in rice (*Oryza sativa L.*) subject to drought during the booting stage occurs not because of lack of assimilate or of water deficit in the shoot but because of dehydration of the root zone. Jpn J Crop Sci 63:510–517

Kobata T, Okuno T, Yamamoto T (1996) Contributions of capacity for soil water extraction and water use efficiency to maintenance of dry matter production in rice subjected to drought. Jpn J Crop Sci 65:652–662

Kobata T, Okuno T, Yamamoto T (1998) Contribution of water use efficiency to crop production of rice subject to soil desiccation. Conference: Asian Crop Science, Taichung, Taiwan, pp. 83–90

Kobata T, Koç M, Barutçular C, Matsumoto T, Nakagawa H, Adachi F, Ünlü M (2012) Assimilate supply as a yield determination factor in spring wheat under high temperature conditions in the Mediterranean zone of south-east Turkey. Plant Prod Sci 15:216–227

Kobata T, Palta JA, Tanaka T, Ohnishi M, Maeda M, KoÇ M, Barutçulare G (2018a) Responses of grain filling in spring wheat and temperate-zone rice to temperature: similarities and differences. Field Crop Res 215:187–199

Kobata T, KoÇ M, Barutçulare G, Tanno K, Inagaki M (2018b) Harvest index is a critical factor influencing the grain yield of diverse wheat species under rain-fed conditions in the Mediterranean zone of southeastern Turkey and northern Syria. Plant Prod Sci. 21:71–81

Legg OJ, Meisinger JJ (1982) Soil nitrogen budgets. Nitrogen Agric Soils 22:503–566

Loomis RS, Connor DJ (1992) Crop ecology: productivity and management in agricultural systems. Cambridge University Press, Cambridge, pp 32–59

Marschner H (1995) Mineral nutrition of higher plants, 2ndedn edn. Academic Press, London

Matsui M, Hasegawa T (2019) Efect of long anther dehiscence on seed set at high temperatures during flowering in rice (*Oryza sativa L.*). Sci Rep 9:20363 |. https://doi.org/10.1038/s41598-019-56792-2

Mian MAR, Nafziger ED, Kolb FL, Teyker RH (1994) Root size and distribution of fieldgrown wheat genotypes. Crop Sci 34:810–812

Nagato Y (1949) On the effects of wilting upon the maturation of rice kernels. Jpn J Crop Sci. 18:7

Nakagawa H, Kobata T, Yano T, Barutcular C, Koc M, Tanaka K, Nagano T, Fujiwara Y, Hoshikawa K, Kume T, Watanabe T (2006) Predicting the impact of global warming on wheat production in Adana. ICCAP Project. https://www.chikyu.ac.jp/iccap/ICCAP_Final_Report/5/3-crop_nakagawa.pdf

Nielsen RL (2015) Effects of flooding or ponding on corn prior to tasseling. http://www.kingcorn.org/news/timeless/pondingyoungcorn.html

Nuttall JG, Barlow KM, Delahunty AJ, Christy BP, O'Leary GJ (2018) Acute high temperature response in wheat. Agron J 110:1296–1308

O'Toole JC, Bland W (1987) Genotypic variation in crop plant root systems. J Adv Agron 41:91–145

Olgun M, Kumlay AM, Adiguzel MC, Caglar A (2008) The effect of waterlogging in wheat (*T. aestivum L.*). Acta Agric Scand Sect B Soil Plant Sci 58:193–198

Panda D, Mishra SS, Behera PK (2021) Review drought tolerance in Rice: focus on recent mechanisms and approaches. Rice Sci 28:119–132

Passioura J (1977) Grain yield, harvest index, and water use of wheat. J Aust Inst Agric Sci 43:117–120

Poorter H, Bühler J, van Dusschoten D, Climent J, Postma JA (2012) Pot size matters: a meta-analysis of the effects of rooting volume on plant growth. Func Plant Biol 39:839–850

RafaelSánchez-Rodríguez A, Nie C, Hill PW, Chadwick DR, Jones DL (2019) Extreme flood events at higher temperatures exacerbate the loss of soil functionality and trace gas emissions in grassland. Soil Biol Biochem 130:227–236

Ray JD, Sinclair TR (1998) The effect of pot size on growth and transpiration of maize and soybean during water deficit stress. J Exp Bot 49:1381–1386

Sadras VS, McDonald G (2012) Water use efficiency of grain crops in Australia: principles, benchmarks and management GRDC project code: DAS00089. https://www.grdc.com.au/~/media/documents/resources/publications/grdcpublicationwateruseefficiencyofgraincropsinaustraliapdf.pdf

Sakagami J, Sone C, Nakazono M (2012) Injury to rice plants by floods and resistance to submergence. Jpn J Crop Sci 81:1–9

Sharma DL, D'Antuono MF, Anderson WK, Shakley BJ, Zaicou-Kunesch CM, Amjad M (2008) Variability of optimum sowing time for wheat yield in West. Aust J Agric Res 59:958–970

Shortridge J (2019) Observed trends in daily rainfall variability result in more severe climate change impacts to agriculture. Clim Chang 157:294–444

Sinclair TR, Tanner CB, Bennett HM (1984) Water-use efficiency in crop production. Bioscience 34:36–40

Suge H (1987) Occurrence of internodal elongation ability under submergence in Japanese native rice varieties, (*Oryza sativa L.*). Jpn J Crop Sci 56:257–262

Tausz-Posch S, Tausz M, Bourgault M (2020) Advances in understanding acclimation, nitrogen dynamics and interactions with drought and other organisms. Plant Biol 22:1–152

Tazaki T, Ishihara K, Ushijima T (1980) Influence of water stress on the photosynthesis and productivity of plants in humid areas. In: Turner NC, Kramer PJ (eds) Adaptation of plants to water and high temperature stress. John Wiley & Sons, Inc, pp 309–339

Tisdall JM (1994) Possible role of soil microorganisms in aggregation in soils. Plant Soil 159:115–121

Turner NC (1986) Crop water deficits: a decade of progress. Adv Agron 39:1–51

Turner NC (1996) Future progress in crop water relations. Adv Agron 58:293–338

Turner NC (2019) Imposing and maintaining soil water deficits in drought studies in pots. Plant Soil 439:1–11

Turner NC, Blum A, Cakir M, Steduto P, Roberto D, Tuberos R, Young N (2014) Strategies to increase the yield and yield stability of crops under drought – are we making progress? Func Plant Biolo 41:1199–1206

Ulfat A, Shokat S, Li X, Fang L, Großkinsky DK, Majid SA, Roitsch T, Liu F (2021) Elevated carbon dioxide alleviates the negative impact of drought on wheat by modulating plant metabolism and physiology. Agr Water Manag 250:106804

Unkovich M, Baldock J, Forbes M (2010) Variability in harvest index of grain crops and potential significance for carbon accounting: examples from Australian agriculture. Adv Agron 105:173–219

Wang B, Cai W, Li J, Wan Y, Li Y, Guo C, Wilkes A, You S, Gao Q, Liu K (2020) Leaf photosynthesis and stomatal conductance acclimate to elevated [CO_2] and temperature thus increasing dry matter productivity in a double rice cropping system. Field Crop Res 248:107735

Wasko C, Nathan R, Stein L, O'Shea D (2021) Evidence of shorter more extreme rainfalls and increased flood variability under climate change. J Hydro 603:126994

Wu JD, Li JC, Wei FZ, Wang CY, Zhang Y, Sun G (2014) Effects of nitrogen spraying on the post-anthesis stage of winter wheat under waterlogging stress. Acta Physiol Plantrum 36:207–216

Wuebker EF, Mullen RE, Koehler K (2001) Flooding and temperature effects on soybean germination. Crop Sci 41:1857–1861

Part II
Root Responses

Chapter 2
Function and Regulation of Aquatic Adventitious Roots

Ole Pedersen, Dan Liu, Lucas León Peralta Ogorek, Margret Sauter, and Chen Lin

2.1 Aquatic Adventitious Roots

Numerous species of both monocots and dicots possess the ability to grow adventitious roots from the stem as a response to flooding. Roots start emerging only a few days after the submergence kicks in (Inouye and Mochizuki 1980), and the swift response to flooding is required since the existing root system will be exposed to soil anoxia soon after waterlogging occurs (Ponnamperuma 1972). Therefore, a new root system of high tissue porosity is needed to sustain root function (Steffens and Rasmussen 2016). If the flooding is restricted to waterlogging, the root emergence occurs from the lower part of the stem, and the new roots penetrate into the wet soil

O. Pedersen (✉)
Department of Biology, University of Copenhagen, Copenhagen, Denmark

School of Biological Sciences, The University of Western Australia, Crawley, WA, Australia
e-mail: opedersen@bio.ku.d

D. Liu
Department of Biology, University of Copenhagen, Copenhagen, Denmark

Plant Developmental Biology and Plant Physiology, University of Kiel, Kiel, Germany

L. L. P. Ogorek
Department of Biology, University of Copenhagen, Copenhagen, Denmark

M. Sauter
Plant Developmental Biology and Plant Physiology, University of Kiel, Kiel, Germany

C. Lin (✉)
Plant Developmental Biology and Plant Physiology, University of Kiel, Kiel, Germany

Key Laboratory of Plant Functional Genomics of the Ministry of Education, Yangzhou University, Yangzhou, Jiangsu, China
e-mail: clin@yzu.edu.cn

J.-I. Sakagami, M. Nakazono (eds.), *Responses of Plants to Soil Flooding*,
https://doi.org/10.1007/978-981-99-9112-9_2

(Sauter 2013). However, under partial or complete submergence, root emergence can occur from further up the stem, and these roots may never reach the soil. Instead, they form a curtain composed of numerous roots floating in the floodwater (Fig. 2.1), and such roots are referred to as aquatic adventitious roots (Rich et al. 2011).

In deepwater rice, the basal soil root system perishes after months of partial submergence (Catling 1992). Loss of function including attachment of the basal root system is supported by observations of easy uprooting during storms but yield is hardly affected by such uprooting (Catling 1992). There is only anecdotal evidence that dieback of the basal roots is caused by soil anoxia (Colmer et al. 2014), but it is likely that the basal soil roots are not of sufficiently high tissue porosity to sustain a diffusive flux of molecular O_2 to the root tips and/or these roots lack a barrier to radial oxygen loss; both would result in dysfunctional roots, but not necessarily in root death per se. Nevertheless, the formation of aquatic adventitious roots is impressive with examples of more than 225 roots per stem in deepwater rice (*cv.* Habiganj Aman VIII) following a partial submergence of 10 weeks (Inouye and Mochizuki 1980) or in *Solanum dulcamara* where the mass of the aquatic adventitious roots can reach *c.* 10% of total plant mass following 5 weeks of partial submergence (Zhang et al. 2015).

Aquatic adventitious can be formed by a number of crops and wild species (Table 2.1). However, aquatic adventitious roots have been most thoroughly studied in monocot species with emphasis on the model plant deepwater rice (*Oryza sativa*)

Fig. 2.1 In situ photo of aquatic adventitious roots in *Meionectes brownii* (*A*) and root cross-section showing chloroplasts in the cortex (*B*). In *A*, the white arrows point at three aquatic adventitious roots, and in *B*, the arrow points at chloroplasts in the cortex visualized by autofluorescence. Scale bar in *A* = 10 mm and in *B* = 50 μm. (*A*) photo by Ole Pedersen; (*B*) cross-section from Rich et al. (2011)

Table 2.1 Examples of species with aquatic adventitious roots

Species name	Mono- or dicots	Chloroplasts[a]	Reference(s)
Acorus calamus	Monocot	Yes	Armstrong and Armstrong (1994)
Oryza sativa	Monocot	Yes	Armstrong and Armstrong (1994)
Phragmites australis	Monocot	Yes	Armstrong and Armstrong (1994)
Zea mays	Monocot	Yes	Armstrong and Armstrong (1994)
Alternanthera philoxeroides	Dicot	No	Ayi et al. (2016)
Cotula coronopifolia	Dicot	Yes	Rich et al. (2012)
Epilobium hirsutum	Dicot	Yes	Armstrong and Armstrong (1994)
Eucalyptus camaldulensis	Dicot	No	Argus et al. (2015)
Helianthus annuus	Dicot	No	Wample and Reid (1975)
Larix laricina	Dicot	No	Calvo-Polanco et al. (2012)
Meionectes brownii	Dicot	Yes	Rich et al. (2011)
Rumex palustris	Dicot	No	Peeters et al. (2002)
Solanum dulcamara	Dicot	No	Zhang et al. (2015)
Solanum lycopersicum	Dicot	No	Phatak et al. (1981)
Tecticornia pergranulata	Dicot	Yes	Pedersen et al. (2006), Rich et al. (2008)

[a]Studies where chloroplasts were not detected in the aquatic adventitious roots were not necessarily aimed at diagnosing this traits

or near isogenic lines (NILs) with one or more quantitative trait loci (QTLs) introgressed from deepwater rice (Lin et al. 2021; Steffens et al. 2012). Below we discuss the hormonal regulation of root emergence as well as the physiological functions of aquatic adventitious roots.

2.2 Genetic and Hormonal Regulation of Root Emergence

Aquatic adventitious roots either develop from dormant pre-formed primordia in the stem or de novo as a response to flooding. There is a large body of literature on the roles of phytohormones linking the formation of adventitious roots showing that auxin, cytokinins, and ethylene in particular serve important regulatory roles (Geiss et al. 2009; Lakehal and Bellini 2019). However, most of the research referred to below under de novo formation is based upon adventitious roots in general, and therefore we refer to adventitious roots rather than aquatic adventitious roots in the following.

2.2.1 De Novo Formation

Auxin and cytokinins regulate the formation of adventitious roots in an antagonistic manner (de Klerk et al. 1999; Kuroha and Satoh 2007). Auxin is indispensable for the formation of adventitious roots, which is controlled at four different levels: (i) biosynthesis, (ii) transport, (iii) conjunction, and (iv) degradation (da Costa et al. 2013). In higher plants, auxin is synthesized via a highly conserved YUC/TAA pathway. Tryptophan (Trp) is converted to indole-3-pyruvate (IPA) through aminotransferase (TAA) as shown in *Arabidopsis thaliana* (Stepanova et al. 2008). Subsequently, YUCCA flavin monooxygenases trigger IPA to form indole-3-acetic acid (IAA) via oxidative decarboxylation. Consequently, the lack of *TAA1* in rice plants leads to longer seminal roots and fewer adventitious roots (Zhang et al. 2018), whereas overexpression of several *YUCCA* genes substantially enhanced adventitious root formation (Yamamoto et al. 2007).

Polar auxin transport is a prerequisite for adventitious root initiation. PIN-formed (PIN) proteins are auxin efflux carriers, and the majority of *PIN* genes are highly expressed in adventitious roots (Lin and Sauter 2019) and the *pin1* rice mutant showed root growth deficit (Liu et al. 2009). *CROWN ROOTLESS 1/ADVENTITIOUS ROOTLESS 1* (*CRL1/ARL1*) and *CROWN ROOTLESS 5* (*CRL5*) regulate the number of adventitious roots independently in an auxin-dependent manner, further emphasizing the regulatory role of auxin in adventitious root formation (Inukai et al. 2005; Kitomi et al. 2011; Liu et al. 2005). Both *CRL1* and *CRL5* are direct target genes of auxin response factors (ARFs). Deletion of either one of these two genes caused a severe reduction in the number of roots in rice. In addition, *CRL5* promotes adventitious root initiation through positively activating cytokinin type-A response regulator (type-A RR), OsRR1, thus repressing cytokinin signalling and suggesting the role of *CRL5* in integrating auxin and cytokinin signalling pathways (Kitomi et al. 2011). Rice cytokinin oxidase/dehydrogenase 4 (*OsCKX4*) is a dominant gene positively controlling adventitious root numbers in rice. *OsCKX4* is activated by the auxin response factor OsARF25 and the cytokinin response regulators OsRR2 and OsRR3, indicating its role as an integrator to mediate auxin and cytokinin signalling. The expression of *OsCKX4* is also induced by exogenously added auxin and cytokinins (Gao et al. 2014). The *WUSCHEL*-related homeobox (*WOX*) gene, *WOX11*, is seen as a root signature gene in rice, and it is expressed not only in emerging adventitious roots but also in the cell division regions of the root meristem (Zhao et al. 2015, 2009). Similarly to *OsCKX4*, the expression of *WOX11* is also highly induced by exogenously added auxin or cytokinins. Overexpression of *WOX11* in rice resulted in ectopic additional root growth even in non-submerged conditions, and knockdown or deletion of *WOX11* in rice showed fewer roots, thus supporting its critical role in adventitious root development. WOX11 interacts with ethylene response factor 3 (ERF3), and they both target the cytokinin-responsive gene, *RR2*, to regulate the root growth in a cytokinin-dependent manner (Zhao et al. 2015). Moreover, WOX11 also recruits ADA2/GCN5 histone acetyltransferase module to activate the downstream gene expression related to cell proliferation in the root meristem, suggesting the involvement of epigenetic regulation in adventitious root formation (Zhou et al. 2017).

In addition to auxin and cytokinins, some other phytohormones participate in the development of adventitious roots. These are ethylene (Chen et al. 2018; Li et al. 2015; Merchante et al. 2013) with additional involvement of strigolactones (Arite et al. 2012), abscisic acid (Xiao et al. 2016), and brassinosteroids (Mori et al. 2002).

Much research has suggested that the increased level of ethylene and strigolactones alter auxin levels, indicating a strong interaction between the different phytohormones (Hu et al. 2018; Robles et al. 2013). However, our understanding of adventitious root development orchestrated by different phytohormones is still limited. Consequently, the extent to which these phytohormones control the initiation of adventitious roots independently or rely on changes in the levels of auxin or cytokinins needs to be investigated.

2.2.2 *Growth of Aquatic Adventitious Roots from Pre-formed Primordia*

Growth of aquatic adventitious roots from pre-formed primordia generally occurs as a response to flooding. The emergence of aquatic adventitious roots in deepwater rice is also affected by light, water temperature, water depth, and the age of the node (Lorbiecke and Sauter 1999). Emergence and growth of aquatic adventitious roots are complex processes combining cell division in the root apical meristem and elongation of the basal part of the root (Lorbiecke and Sauter 1999; Steffens and Rasmussen 2016).

In deepwater rice, the dormant, pre-formed root primordia can resume growth as a response to flooding. Following submergence of the stem, constitutively produced ethylene accumulates in the tissue since radial diffusion out of the stem is restricted by the slow diffusion of gases in water, and the increase in tissue ethylene further stimulates ethylene production (Fig. 2.2) (Bailey-Serres and Voesenek 2010). Flooding induces ethylene biosynthesis via the rate-limiting enzyme 1-aminocyclo propane-1-carboxylic acid synthase in both rice and tomato (Vidoz et al. 2010), and due to increased tissue concentrations of ethylene, the root primordia resume growth (Lin and Sauter 2019). The key role of ethylene can be demonstrated by chemical application of ethylene perception inhibitor 1-methylcyclopropene (1-MCP), arresting the penetration of aquatic adventitious roots in rice (Lin and Sauter 2018) as well as in cucumber (Qi et al. 2019).

In rice, the root primordia are covered by epidermal cells, and therefore the young aquatic adventitious roots initially need to penetrate the epidermal layer before further root elongation can occur unhindered (Bleecker et al. 1986; Lin and Sauter 2018, 2019). In deepwater rice, the actual penetration of the epidermis is facilitated via coordinated programmed cell death (Fig. 2.2) (Steffens et al. 2012). Following the induction of ethylene biosynthesis, a mechanical force is also needed to penetrate the epidermal cells so that aquatic adventitious roots are eventually exposed to the surrounding floodwaters (Lin and Sauter 2019). Ethylene induces the penetration and growth of aquatic adventitious roots together with gibberellin acid in a synergistic manner. In contrast, gibberellin acid alone does not result in outgrowth of roots, and abscisic acid inhibited the ethylene-promoting effect of aquatic adventitious root penetration in deepwater rice (Steffens et al. 2006). We therefore

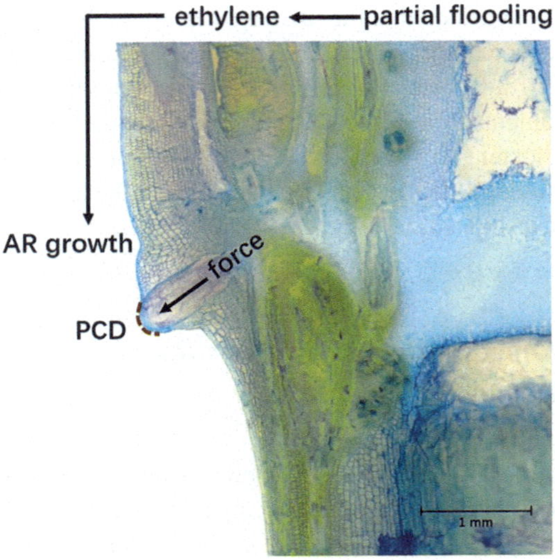

Fig. 2.2 Conceptual model showing the outgrowth of adventitious roots as a response to submergence. Partial flooding (i.e., still with a portion of the shoot in air but with a proportion of the stem submerged) leads to accumulation of ethylene in the tissue. Accumulation of ethylene releases the dormant root primordia, and the growing root generates sufficient force to penetrate the stem following programmed cell death (PCD) of the epidermal cells. Diagram based on the conceptual model in Lin and Sauter (2020)

propose that the various interactions between these phytohormones serve a role in timing the epidermal cell death and root emergence and to further regulate the growth rate of emerging aquatic adventitious roots.

2.3 Two Types of Aquatic Adventitious Roots in NIL-12

Deepwater rice survives along the banks of rivers and lakes and acclimates to fluctuations in water level through different adaptive traits including stem elongation and adventitious root growth (Catling 1992). Three quantitative trait loci (QTLs, viz., QTL1, QTL3, and QTL12) have been described to regulate stem elongation in deepwater rice *cv.* C9285 (Hattori et al. 2009), and the rapid stem elongation is mediated through ethylene (Métraux and Kende 1983), which is acting through gibberellin acid (Kuroha et al. 2018). The stem elongation enables the plant to keep track with the rising waters, and the "snorkelling" helps acquire O_2 from the atmosphere so that internal aeration is sustained (Mori et al. 2019). Recently, it has also been shown that QTL1 and QTL12 not only control stem elongation but also enhance outgrowth of aquatic adventitious roots during partial submergence in rice (Lin et al. 2023).

In deepwater rice and in the near isogenic line with QTL12 introgressed (NIL-12), two different types of aquatic adventitious root are formed as a response to partial submergence (Inouye and Mochizuki 1980; Lin et al. 2023). Within 2–3 days of submergence, thin and slender aquatic adventitious roots emerge from the node from a point just below the position where the leaf sheath is attached (Fig. 2.3); these roots are referred to as adventitious roots 1 (AR1) (Lin et al. 2023). If the node remains submerged, another set of adventitious roots emerge, which are thicker and emerge from a point where they penetrate the stem as well as the most basal part of the leaf sheath (Fig. 2.3); these roots are referred to as adventitious roots 2 (AR2) (Lin et al. 2023). AR1 and AR2 differ morphologically and anatomically indicating distinct functions. The combined use of O_2 microsensors and an apoplastic tracer revealed that AR2 form a barrier to radial O_2 loss (ROL), whereas AR1 do not (Lin et al. 2023). Thin roots without a ROL barrier desiccate faster than thick roots with a barrier (Peralta Ogorek et al. 2021; Song et al. 2022), and thus AR1 desiccate faster than AR2 when exposed to a dry atmosphere (Fig. 2.3). Moreover, the cortex to stele ratio also indicates better adaptation of AR2 to low water availability than in AR1 roots (Lin et al. 2023). Intriguingly, the mechanical impedance provided by the leaf sheath is a key factor for AR2 thickness; once the leaf sheath is removed, the AR2 become comparatively thinner after penetration (Lin et al. 2023). The maximum root length of AR1 and AR2 also differs with AR1 reaching *c.* 20 cm and AR2 *c.* 35 cm after 60 days of partial submergence (Lin et al. 2023).

The eco-physiological importance of these traits in a real flood situation where the water may rise and fall multiple times during the wet season is, however, questionable. Both types of roots perish soon after de-submergence when exposed to dry air (Lin et al. 2023). However, in a natural environment, the stems of deepwater would not remain erect with the roots exposed to air but instead follow the water level as it recedes (see figure in Catling (1992)). The entangled shoots and numerous adventitious roots would create their own microclimate with much higher relative humidity partly protecting the roots from desiccation, and some roots would continue growing and find their way into the moist soil. Nevertheless, if the aquatic adventitious roots formed during the first flood event die, the growth potential of AR1 is very limited as most of the primordia had been released during the first flood event (Lin et al. 2023). On the other hand, the growth potential of AR2 still remains, and a new set of AR2 are formed as a response to a second flood event (Lin et al. 2023), suggesting that AR2 are likely ecologically more important than AR1. We therefore propose that AR2 roots may not only grow from pre-formed primordia but also form de novo. Moreover, the importance of AR2, in particular, for water and nutrient uptake during partial submergence should be further explored to justify introducing aquatic adventitious roots in modern cultivars for areas experiencing long-term shallow flooding (Lin et al. 2023).

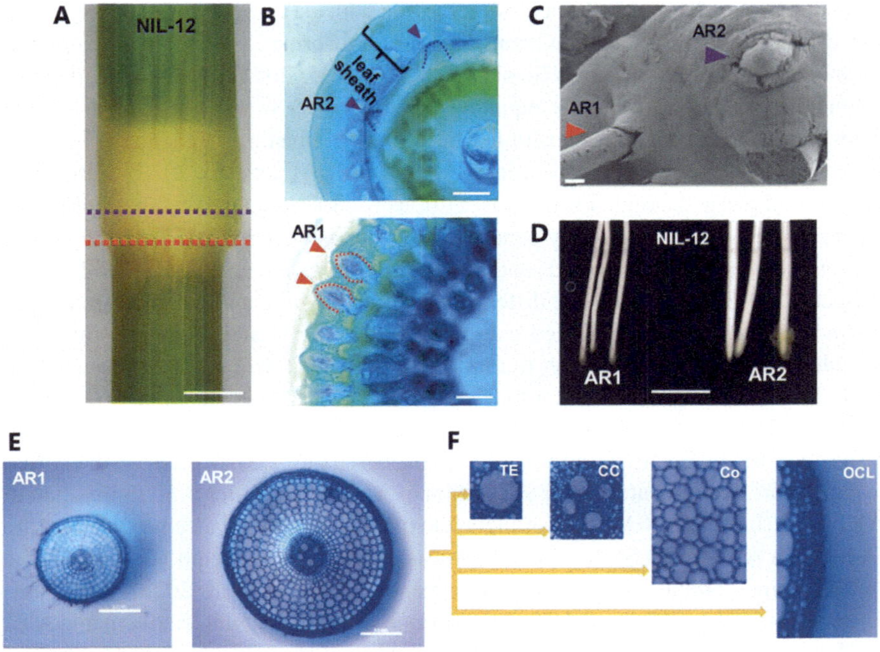

Fig. 2.3 NIL-12 is a near isogenic line of paddy rice (*cv.* Taichung 65) introgressed with a QTL (QTL12) from deepwater rice encoding stem elongation as a response to flooding (Hattori et al. 2009). NIL-12 forms two distinct types of aquatic adventitious roots at the nodes. (**a**) NIL-12 stem node indicating the two positions at which the two types of aquatic adventitious roots develop (a red line from where AR1 first grow and purple line from where AR2 emerge a few days later). Scale bar = 5 mm. (**b**) Nodal cross-sections at the positions indicated in A. Note that AR1 only has to penetrate the epidermis, whereas AR2 need to penetrate the epidermis as well as the basal part of the leaf sheath. Scale bar = 1 mm. (**c**) Scanning electron micrograph showing emerging AR1 and AR2. Scale bar = 300 µm. (**d**) Examples of AR1 and AR2. Scale bar = 5 mm. (**e**) Cross-sections of AR1 and AR2 showing distinct differences in root diameter due to contrasts in both cortical and stelar dimensions. Scale bar = 200 µm. (**f**) Tracheary elements (TE), central cylinder (CC), cortex (Co), and outer cell layers (OCL) are shown for AR2. The cross sections were taken 5 mm behind the root tip. Scale bar = 200 µm. Figure from Lin et al. (2023)

2.4 Nutrient Uptake by Aquatic Adventitious Roots

Considering that a wide range of monocots and dicots form aquatic adventitious roots as a response to flooding, little is known about their role in nutrient uptake. A study on deepwater rice using stable inorganic nitrogen isotopes (^{15}N) to distinguish between N uptake by the basal soil roots and aquatic adventitious roots showed that the aquatic adventitious roots were the main sites for inorganic N uptake (Khan et al. 1982). Moreover, when ^{15}N was supplied to the water prior to flowering, the main component of grain nitrogen was derived from the floodwater via uptake by the aquatic adventitious roots (Khan et al. 1982). Similarly for inorganic phosphorus (P), the aquatic adventitious roots were the main sites for P uptake in *Nasturtium officinale* despite a much larger proportion of soil roots (Cumbus and Robinson 1977). However, some studies using isotopes have a tendency to overestimate the importance of the aquatic adventitious since the concentrations at which the isotopes are supplied are sometimes unrealistically high.

A more recent study of *Solanum dulcamara* forming numerous aquatic adventitious roots as a response to partial submergence showed strong, positive correlation between uptake of phosphate from the floodwater and total mass of aquatic adventitious roots (Zhang et al. 2017a). Moreover, plants with numerous aquatic adventitious roots had higher transpiration rates and faster growth than plants where the aquatic adventitious roots were experimentally removed (Zhang et al. 2017a). The benefits of aquatic adventitious roots only appeared during prolonged partial submergence, and the growth of these roots ceased under complete submergence. Plant carbohydrate status also influenced growth of aquatic adventitious roots in *S. dulcamara*, but poor carbohydrate status prior to flooding could be partly alleviated through high light during the submergence event (Zhang et al. 2017b) possibly due to higher underwater net photosynthesis (Colmer et al. 2011) and thereby de novo carbohydrate production.

2.5 Oxygen Exchange by Aquatic Adventitious Roots

Aquatic adventitious roots can act as "gills" during flooding facilitating O_2 exchange with the surrounding water. Completely submerged *Alternanthera philoxeroides* with aquatic adventitious roots maintained a threefold higher stem tissue O_2 status in darkness than stems where the roots had been experimentally removed (Ayi et al. 2016). Pending that the roots do not form a barrier to ROL as some roots of deepwater rice do (Lin et al. 2021), the aquatic adventitious roots are likely more gas permeable than stem tissues, but more importantly the large surface area of the roots facilitates gas exchange between the floodwater and the plant body. Since the adventitious roots are well connected with the stem via aerenchyma, O_2 entering from the floodwater can diffuse unhindered into the stem (Ayi et al. 2016; Rich et al. 2013). However, when partially submerged, it is expected that the aquatic adventitious roots rely on O_2 supply from the stem as in the case of deepwater rice (Lin et al. 2021) or if the roots lack chloroplasts (see next section).

During the diurnal cycle, aquatic adventitious roots can experience huge fluctuations in tissue O_2 status. For example, aquatic adventitious roots of *Tecticornia pergranulata* fluctuated between anoxia in the cortex during darkness followed by O_2 reaching 15 kPa pO_2 only 3 h after sunrise (Pedersen et al. 2006). Similarly, the aquatic adventitious roots of *Meionectes brownii* cycled between low values late in the night (down to 3 kPa pO_2) and extremely high values (43 kPa pO_2) in the early afternoon (Rich et al. 2013). Such dramatic changes are also found in adventitious roots buried in anoxic soils, as demonstrated in completely submerged paddy rice where the cortical O_2 cycled between apparent anoxia and up to 14 kPa pO_2 at midday (Winkel et al. 2013). The influence on root functioning of such dramatic, daily fluctuations in root tissue O_2 status is likely profound. Laboratory experiments involving submerged rice seedlings have demonstrated that root elongation occurred only during the light when O_2 was available and not only ceased during the night as a result of lack of O_2 but also resulted in ethanol production due to fermentation (Waters et al. 1989).

2.6 Green Photosynthetic Aquatic Adventitious Roots

Several aquatic adventitious roots develop fully functional chloroplasts so that the roots contribute to the carbon balance and supply molecular O_2 when in light. Some examples are from light-exposed aquatic adventitious roots of *Acorus calamus*, *Epitohium hirsutum*, *Phragmites australis*, maize, and rice. Here, chloroplasts developed in the inner and outermost sections of non-aerenchymatous cortical tissues and even in the radial cell strands between the gas-filled lacunae (Armstrong and Armstrong 1994).

Chloroplasts forming in the aquatic adventitious roots of *Tecticornia pergranulata* were found to be fully functional. Immunodetection confirmed the presence of ribulose-1-5-bisphosphate carboxylase/oxygenase (Rubisco) in the chloroplasts, and the chloroplasts contained chlorophyll$_a$ (resulting in the autofluorescence), although the concentration was only 6% of that in the stem (Rich et al. 2008). In contrast, chlorophyll$_b$ showed much higher concentrations in the aquatic adventitious roots, being 36% of that in the stem, demonstrating acclimation to the low light environment within the water (Rich et al. 2008). The green photosynthetic roots showed a saturating response in underwater net photosynthesis to increasing light at *c.* 1000 μmol photons m^{-2} s^{-1} and with a light compensation point (I_C) at *c.* 250 μmol photons m^{-2} s^{-1} (Rich et al. 2008). Similarly, photosynthetic O_2 evolution has been documented in *Meionectes brownii* as cortical pO_2 increased quickly when the roots were illuminated (Rich et al. 2011); see Fig. 2.1b for examples of autofluorescence from chloroplasts in the cortex of *Meionectes brownii*. The study of Rich et al. (2011) demonstrated the presence of a complete photosynthetic pathway, capable of both O_2 evolution and carbon fixation (visualized by ^{14}C) within aquatic adventitious roots of a semi-aquatic species. With the abundant growth of an extensive aquatic root system, the photosynthetic ability likely offers an advantage to submerged plants as these roots need little O_2 and carbohydrate inputs from the shoot during the day, which is in stark contrast to soil roots that rely 100% on molecular O_2 and carbohydrates from the shoot.

2.7 Conclusion and Outlook

Several species of wild wetland plants and economically important crops are known to form adventitious roots as a response to flooding. Roots that do not reach the soil surface are referred to as aquatic adventitious roots, and the outgrowth of these roots is controlled by ethylene if emergence occur from pre-formed root primordia. During de novo formation, auxin and cytokinins regulate root growth in an antagonistic manner, and root penetration of the epidermal cells of the stem follows programmed cell death to pave the road for the emerging roots. Deepwater rice forms two distinct types of aquatic adventitious root, viz., AR1 and AR2, but we propose that AR2 are more important than AR1 during long-term flood events. Aquatic adventitious roots take up inorganic N and P from the floodwater and gradually replace the function of soil roots. Some aquatic adventitious roots can develop fully functional chloroplasts and thereby contribute to the O_2 and carbon balance of the flooded plants.

Considering the important role of aquatic adventitious roots during partial submergence, little research has been conducted to further advance our understanding of these roots particularly under field conditions. To a certain extent, paddy rice can also form aquatic adventitious roots as a response to partial submergence. We therefore propose that the role of aquatic adventitious roots for water and nutrient uptake should be further studied in order to uncover their breeding potential for developing new rice cultivars for flood-prone areas.

References

Argus RE, Colmer TD, Grierson PF (2015) Early physiological flood tolerance is followed by slow post-flooding root recovery in the dryland riparian tree *Eucalyptus camaldulensis* subsp. *refulgens*. Plant Cell Environ 38:1189–1199. https://doi.org/10.1111/pce.12473

Arite T, Kameoka H, Kyozuka J (2012) Strigolactone positively controls crown root elongation in rice. J Plant Growth Regul 31:165–172. https://doi.org/10.1007/s00344-011-9228-6

Armstrong J, Armstrong W (1994) Chlorophyll development in mature lysigenous and schizogenous root aerenchymas provides evidence of continuing cortical cell viability. New Phytol 126:493–497. https://doi.org/10.1111/j.1469-8137.1994.tb04246.x

Ayi Q, Zeng B, Liu J, Li S, van Bodegom PM, Cornelissen JHC (2016) Oxygen absorption by adventitious roots promotes the survival of completely submerged terrestrial plants. Ann Bot 118:675–683. https://doi.org/10.1093/aob/mcw051

Bailey-Serres J, Voesenek LACJ (2010) Life in the balance: a signaling network controlling survival of flooding. Curr Opin Plant Biol 13:489–494. https://doi.org/10.1016/j.pbi.2010.08.002

Bleecker AB, Schuette JL, Kende H (1986) Anatomical analysis of growth and developmental patterns in the internode of deepwater rice. Planta 169:490–497. https://doi.org/10.1007/BF00392097

Calvo-Polanco M, Señorans J, Zwiazek JJ (2012) Role of adventitious roots in water relations of tamarack (*Larix laricina*) seedlings exposed to flooding. BMC Plant Biol 12:1–9. https://doi.org/10.1186/1471-2229-12-99

Catling D (1992) Rice in deep water. International Rice Research Institute, Los Baños, The Philippines

Chen H, Ma B, Zhou Y, He S-J, Tang S-Y, Lu X, Xie Q, Chen S-Y, Zhang J-S (2018) E3 ubiquitin ligase SOR1 regulates ethylene response in rice root by modulating stability of Aux/IAA protein. Proc Natl Acad Sci 115:4513–4518. https://doi.org/10.1073/pnas.1719387115

Colmer TD, Winkel A, Pedersen O (2011) A perspective on underwater photosynthesis in submerged terrestrial wetland plants. AoB Plants 2011:plr030. https://doi.org/10.1093/aobpla/plr030

Colmer TD, Armstrong W, Greenway H, Ismail AM, Kirk GJD, Atwell BJ (2014) Physiological mechanisms of flooding tolerance in rice: transient complete submergence and prolonged standing water. Prog Bot 75:255–307. https://doi.org/10.1007/978-3-642-38797-5_9

Cumbus IP, Robinson LW (1977) The function of root systems in mineral nutrition of watercress (*Rorippa nasturtium-aquaticum* (L) Hayek). Plant Soil 47:395–406

da Costa CT, de Almeida MR, Ruedell CM, Schwambach J, Maraschin FS, Fett-Neto AG (2013) When stress and development go hand in hand: main hormonal controls of adventitious rooting in cuttings. Front Plant Sci 4:133. https://doi.org/10.3389/fpls.2013.00133

de Klerk G-J, van Der Krieken W, de Jong JC (1999) Review the formation of adventitious roots: new concepts, new possibilities. In Vitro Cell Dev Biol Plant 35:189–199. https://doi.org/10.1007/s11627-999-0076-z

Gao S, Fang J, Xu F, Wang W, Sun X, Chu J, Cai B, Feng Y, Chu C (2014) *CYTOKININ OXIDASE/DEHYDROGENASE4* integrates cytokinin and auxin signaling to control rice crown root formation. Plant Physiol 165:1035–1046. https://doi.org/10.1104/pp.114.238584

Geiss G, Gutierrez L, Bellini C (2009) Adventitious root formation: new insights and perspectives. Ann Plant Rev 37:127–156. https://doi.org/10.1002/9781444310023.ch5

Hattori Y, Nagai K, Furukawa S, Song X-J, Kawano R, Sakakibara H, Wu J, Matsumoto T, Yoshimura A, Kitano H, Matsuoka M, Mori H, Ashikari M (2009) The ethylene response factors *SNORKEL1* and *SNORKEL2* allow rice to adapt to deep water. Nature 460:1026–1030. https://doi.org/10.1038/nature08258

Hu Q, Zhang S, Huang B (2018) Strigolactones and interaction with auxin regulating root elongation in tall fescue under different temperature regimes. Plant Sci 271:34–39. https://doi.org/10.1016/j.plantsci.2018.03.008

Inouye J, Mochizuki T (1980) Emergence of crown roots from the elongated culm in several floating rice varieties under submerged conditions. Jpn J Trop Agric 24(3):125–131

Inukai Y, Sakamoto T, Ueguchi-Tanaka M, Shibata Y, Gomi K, Umemura I, Hasegawa Y, Ashikari M, Kitano H, Matsuoka M (2005) *Crown rootless1*, which is essential for crown root formation in rice, is a target of an AUXIN RESPONSE FACTOR in auxin signaling. Plant Cell 17:1387–1396. https://doi.org/10.1105/tpc.105.030981

Khan M, Ventura W, Vergara B (1982) Uptake through aquatic roots and distribution of ^{15}N-tagged ammonium in Deepwater rice. Proceedings of the 1981 international Deepwater rice workshop IRRI, Philippines. International Rice Research Institute, Los Baños, The Philippines

Kitomi Y, Ito H, Hobo T, Aya K, Kitano H, Inukai Y (2011) The auxin responsive AP2/ERF transcription factor *CROWN ROOTLESS5* is involved in crown root initiation in rice through the induction of OsRR1, a type-A response regulator of cytokinin signaling. Plant J 67:472–484. https://doi.org/10.1111/j.1365-313X.2011.04610.x

Kuroha T, Satoh S (2007) Involvement of cytokinins in adventitious and lateral root formation. Plant Root 1:27–33. https://doi.org/10.3117/plantroot.1.27

Kuroha T, Nagai K, Gamuyao R, Wang DR, Furuta T, Nakamori M, Kitaoka T, Adachi K, Minami A, Mori Y, Mashiguchi K, Seto Y, Yamaguchi S, Kojima M, Sakakibara H, Wu J, Ebana K, Mitsuda N, Ohme-Takagi M, Yanagisawa S, Yamasaki M, Yokoyama R, Nishitani K, Mochizuki T, Tamiya G, McCouch SR, Ashikari M (2018) Ethylene-gibberellin signaling underlies adaptation of rice to periodic flooding. Science 361:181–186. https://doi.org/10.1126/science.aat1577

Lakehal A, Bellini C (2019) Control of adventitious root formation: insights into synergistic and antagonistic hormonal interactions. Physiol Plant 165:90–100. https://doi.org/10.1111/ppl.12823

Li W, Ma M, Feng Y, Li H, Wang Y, Ma Y, Li M, An F, Guo H (2015) EIN2-directed translational regulation of ethylene signaling in Arabidopsis. Cell 163:670–683. https://doi.org/10.1016/j.cell.2015.09.037

Lin C, Sauter M (2018) Control of adventitious root architecture in rice by darkness, light, and gravity. Plant Physiol 176:1352–1364. https://doi.org/10.1104/pp.17.01540

Lin C, Sauter M (2019) Polar auxin transport determines adventitious root emergence and growth in rice. Front Plant Sci 10:444. https://doi.org/10.3389/fpls.2019.00444

Lin C, Sauter M (2020) Control of root system architecture by phytohormones and environmental signals in rice. Israel J Plant Sci 67:98–109. https://doi.org/10.1163/22238980-20191108

Lin C, Peralta Ogorek LL, Pedersen O, Sauter M (2021) Oxygen in the air and oxygen dissolved in the floodwater both sustain growth of aquatic adventitious roots in rice. J Exp Bot 72:1879–1890. https://doi.org/10.1093/jxb/eraa542

Lin C, Peralta Ogorek LL, Liu D, Pedersen O, Sauter M (2023) A quantitative trait locus conferring flood tolerance to deepwater rice regulates the formation of two distinct types of aquatic adventitious roots. New Phytol 238:1403–1419. https://doi.org/10.1111/nph.18678

Liu H, Wang S, Yu X, Yu J, He X, Zhang S, Shou H, Wu P (2005) ARL1, a LOB-domain protein required for adventitious root formation in rice. Plant J 43:47–56. https://doi.org/10.1111/j.1365-313X.2005.02434.x

Liu S, Wang J, Wang L, Wang X, Xue Y, Wu P, Shou H (2009) Adventitious root formation in rice requires OsGNOM1 and is mediated by the OsPINs family. Cell Res 19:1110–1119. https://doi.org/10.1038/cr.2009.70

Lorbiecke R, Sauter M (1999) Adventitious root growth and cell-cycle induction in deepwater rice. Plant Physiol 119:21–30. https://doi.org/10.1104/pp.119.1.21

Merchante C, Alonso JM, Stepanova AN (2013) Ethylene signaling: simple ligand, complex regulation. Curr Opin Plant Biol 16:554–560. https://doi.org/10.1016/j.pbi.2013.08.001

Métraux J-P, Kende H (1983) The role of ethylene in the growth response of submerged deep water rice. Plant Physiol 72:441–446. https://doi.org/10.1104/pp.72.2.441

Mori M, Nomura T, Ooka H, Ishizaka M, Yokota T, Sugimoto K, Okabe K, Kajiwara H, Satoh K, Yamamoto K (2002) Isolation and characterization of a rice dwarf mutant with a defect in brassinosteroid biosynthesis. Plant Physiol 130:1152–1161. https://doi.org/10.1104/pp.007179

Mori Y, Kurokawa Y, Koike M, Malik AI, Colmer TD, Ashikari M, Pedersen O, Nagai K (2019) Diel O_2 dynamics in partially and completely submerged deepwater rice: leaf gas films enhance internodal O_2 status, influence gene expression, and accelerate stem elongation for 'snorkelling' during submergence. Plant Cell Physiol 65:973–985. https://doi.org/10.1093/pcp/pcz009

Pedersen O, Vos H, Colmer TD (2006) Oxygen dynamics during submergence in the halophytic stem succulent *Halosarcia pergranulata*. Plant Cell Environ 29:1388–1399. https://doi.org/10.1111/j.1365-3040.2006.01522.x

Peeters AJM, Cox MCH, Benschop JJ, Vreeburg RAM, Bou J, Voesenek LACJ (2002) Submergence research using *Rumex palustris* as a model; looking back and going forward. J Exp Bot 53:391–398. https://doi.org/10.1093/jexbot/53.368.391

Peralta Ogorek LL, Pellegrini E, Pedersen O (2021) Novel functions of the root barrier to radial oxygen loss – radial diffusion resistance to H_2 and water vapour. New Phytol 231:1365–1376. https://doi.org/10.1111/nph.17474

Phatak SC, Jaworski CA, Liptay A (1981) Flowering and adventitious root growth of tomato cultivars as influenced by ethephon. HortScience 16:181–182. https://doi.org/10.21273/HORTSCI.16.2.181

Ponnamperuma FN (1972) The chemistry of submerged soils. Adv Agron 24:29–96. https://doi.org/10.1016/S0065-2113(08)60633-1

Qi X, Li Q, Ma X, Qian C, Wang H, Ren N, Shen C, Huang S, Xu X, Xu Q (2019) Waterlogging-induced adventitious root formation in cucumber is regulated by ethylene and auxin through reactive oxygen species signalling. Plant Cell Environ 42:1458–1470. https://doi.org/10.1111/pce.13504

Rich SM, Ludwig M, Colmer TD (2008) Photosynthesis in aquatic adventitious roots of the halophytic stem-succulent *Tecticornia pergranulata* (formerly *Halosarcia pergranulata*). Plant Cell Environ 31:1007–1016. https://doi.org/10.1111/j.1365-3040.2008.01813.x

Rich SM, Ludwig M, Pedersen O, Colmer TD (2011) Aquatic adventitious roots of the wetland plant *Meionectes brownii* can photosynthesize: implications for root function during flooding. New Phytol 190:311–319. https://doi.org/10.1111/j.1469-8137.2010.03524.x

Rich SM, Ludwig M, Colmer TD (2012) Aquatic adventitious root development in partially and completely submerged wetland plants *Cotula coronopifolia* and *Meionectes brownii*. Ann Bot 110:405–414. https://doi.org/10.1093/aob/mcs051

Rich SM, Pedersen O, Ludwig M, Colmer TD (2013) Shoot atmospheric contact is of little importance to aeration of deeper portions of the wetland plant *Meionectes brownii*; submerged organs mainly acquire O_2 from the water column or produce it endogenously in underwater photosynthesis. Plant Cell Environ 36:213–223. https://doi.org/10.1111/j.1365-3040.2012.02568.x

Robles L, Stepanova A, Alonso J (2013) Molecular mechanisms of ethylene–auxin interaction. Mol Plant 6:1734–1737. https://doi.org/10.1093/mp/sst113

Sauter M (2013) Root responses to flooding. Curr Opin Plant Biol 16:282–286. https://doi.org/10.1016/j.pbi.2013.03.013

Song Z, Zonta F, Peralta Ogorek LL, Bastegaard VK, Herzog M, Pellegrini E, Pedersen O (2022) The quantitative importance of key root traits for radial water loss under low water potential. Plant Soil 482:567–584. https://doi.org/10.1007/s11104-022-05711-y

Steffens B, Rasmussen A (2016) The physiology of adventitious roots. Plant Physiol 170:603–617. https://doi.org/10.1104/pp.15.01360

Steffens B, Wang J, Sauter M (2006) Interactions between ethylene, gibberellin and abscisic acid regulate emergence and growth rate of adventitious roots in Deepwater rice. Planta 223:604–612. https://doi.org/10.1007/s00425-005-0111-1

Steffens B, Kovalev A, Gorb SN, Sauter M (2012) Emerging roots alter epidermal cell fate through mechanical and reactive oxygen species signaling. Plant Cell 24:3296–3306. https://doi.org/10.1105/tpc.112.101790

Stepanova AN, Robertson-Hoyt J, Yun J, Benavente LM, Xie D-Y, Doležal K, Schlereth A, Jürgens G, Alonso JM (2008) *TAA1*-mediated auxin biosynthesis is essential for hormone crosstalk and plant development. Cell 133:177–191. https://doi.org/10.1016/j.cell.2008.01.047

Vidoz ML, Loreti E, Mensuali A, Alpi A, Perata P (2010) Hormonal interplay during adventitious root formation in flooded tomato plants. Plant J 63:551–562. https://doi.org/10.1111/j.1365-313X.2010.04262.x

Wample RL, Reid DM (1975) Effect of aeration on the flood-induced formation of adventitious roots and other changes in sunflower (*Helianthus annuus* L.). Planta 127:263–270. https://doi.org/10.1007/BF00380723

Waters I, Armstrong W, Thomson CJ, Setter TL, Adkins S, Gibbs J, Greenway H (1989) Diurnal changes in radial oxygen loss and ethanol metabolism in roots of submerged and non-submerged rice seedlings. New Phytol 113:439–451

Winkel A, Colmer TD, Ismail AM, Pedersen O (2013) Internal aeration of paddy field rice (*Oryza sativa* L.) during complete submergence – importance of light and floodwater O_2. New Phytol 197:1193–1203. https://doi.org/10.1111/nph.12048

Xiao G, Qin H, Zhou J, Quan R, Lu X, Huang R, Zhang H (2016) OsERF2 controls rice root growth and hormone responses through tuning expression of key genes involved in hormone signaling and sucrose metabolism. Plant Mol Biol 90:293–302. https://doi.org/10.1007/s11103-015-0416-9

Yamamoto Y, Kamiya N, Morinaka Y, Matsuoka M, Sazuka T (2007) Auxin biosynthesis by the *YUCCA* genes in rice. Plant Physiol 143:1362–1371. https://doi.org/10.1104/pp.106.091561

Zhang Q, Visser EJ, de Kroon H, Huber H (2015) Life cycle stage and water depth affect flooding-induced adventitious root formation in the terrestrial species *Solanum dulcamara*. Ann Bot 116:279–290. https://doi.org/10.1093/aob/mcv095

Zhang Q, Huber H, Beljaars SJM, Birnbaum D, de Best S, de Kroon H, Visser EJW (2017a) Benefits of flooding-induced aquatic adventitious roots depend on the duration of submergence: linking plant performance to root functioning. Ann Bot 120:171–180. https://doi.org/10.1093/aob/mcx049

Zhang Q, Huber H, Boerakker JWT, Bosch D, de Kroon H, Visser EJW (2017b) Environmental factors constraining adventitious root formation during flooding of *Solanum dulcamara*. Funct Plant Biol 44:858. https://doi.org/10.1071/FP16357

Zhang T, Li R, Xing J, Yan L, Wang R, Zhao Y (2018) The YUCCA-auxin-WOX11 module controls crown root development in rice. Front Plant Sci 9:523. https://doi.org/10.3389/fpls.2018.00523

Zhao Y, Hu Y, Dai M, Huang L, Zhou D-X (2009) The WUSCHEL-related homeobox gene *WOX11* is required to activate shoot-borne crown root development in rice. Plant Cell 21:736–748. https://doi.org/10.1105/tpc.108.061655

Zhao Y, Cheng S, Song Y, Huang Y, Zhou S, Liu X, Zhou D-X (2015) The interaction between rice ERF3 and WOX11 promotes crown root development by regulating gene expression involved in cytokinin signaling. Plant Cell 27:2469–2483. https://doi.org/10.1105/tpc.15.00227

Zhou S, Jiang W, Long F, Cheng S, Yang W, Zhao Y, Zhou D-X (2017) Rice homeodomain protein WOX11 recruits a histone acetyltransferase complex to establish programs of cell proliferation of crown root meristem. Plant Cell 29:1088–1104. https://doi.org/10.1105/tpc.16.00908

Chapter 3
Root Plasticity for Adaptation and Productivity of Crop Plants Grown Under Various Water Stresses

Mana Kano-Nakata, Shiro Mitsuya, Yoshiaki Inukai, Roel Suralta, Jonathan Niones, Tsubasa Kawai, and Akira Yamauchi

3.1 Water Stresses Triggering Root Plasticity

3.1.1 Anaerobic Conditions Caused by Flooding and/or Waterlogging

Soil environment is heterogeneous in time and space and thus rarely constant or uniform during plant growth, which triggers root plasticity (Suralta et al. 2018a). Moisture is one such soil environmental factors, and too much water in soil due to heavy rain and/or poor drainage causes flooding/waterlogging, which results in hypoxia (inadequate oxygen) or anoxia (no oxygen).

Such soil anaerobic condition is one of the important constraints that reduce the productivity of the major crop plants like wheat (Oyanagi et al. 2004), maize, and

M. Kano-Nakata · Y. Inukai
International Center for Research and Education in Agricultural Sciences, Nagoya University, Nagoya, Japan

S. Mitsuya
Graduate School of Bioagricultural Sciences, Nagoya University, Nagoya, Japan

R. Suralta · J. Niones
Philippine Rice Research Institute, Science City of Muñoz, Muñoz, Philippines

T. Kawai
International Center for Research and Education in Agricultural Sciences, Nagoya University, Nagoya, Japan

Institute of Crop Science, National Agriculture and Food Research Organization, Tsukuba, Japan

A. Yamauchi (✉)
Asian Satellite Campuses Institute, Nagoya University, Nagoya, Japan
e-mail: ayama@agr.nagoya-u.ac.jp

J.-I. Sakagami, M. Nakazono (eds.), *Responses of Plants to Soil Flooding*,
https://doi.org/10.1007/978-981-99-9112-9_3

soybean (Kono et al. 1987). Even for rice, whether it is lowland, rainfed lowland, or upland variety, which is known to grow well under continuous waterlogged conditions, anaerobic conditions can be very stressful to the plants when they are exposed to such conditions right after being exposed to drought (Suralta et al. 2018a). Therefore, the improvement of tolerance to water stress is an important target to increase and stabilize crop production (Chen et al. 2022).

Root plasticity has been reported to be a key plant trait that improves the tolerance and adaptation ability under such anaerobic conditions (Yamauchi et al. 2018; Suralta et al. 2018a). This includes aerenchyma formation (Pedersen et al. 2021; Yamauchi et al. 2021a, b) and nodal root production (Kono et al. 1988; Pedersen et al. 2021; Yamauchi et al. 2021a, b), resulting in an improved root system development and architecture (Galamay et al. 1992; Hayashi et al. 2013) and hydraulic conductance (Hayashi 2014; Watanabe et al. 2020), which will be discussed in later sections in this chapter.

3.1.2 Water Deficit and Oxygen Deficiency Caused by Soil Moisture Fluctuations

Rice plants are often exposed to alternating recurrences of waterlogging (oxygen deficiency) and water deficit caused by soil moisture fluctuations (SMF) due to erratic rainfall patterns, which typically take place in rainfed lowland rice fields (Wade et al. 1998). In addition, the irrigated rice fields employing a water-saving cultivation technology such as alternate wetting and drying (AWD) and aerobic culture may also experience SMF due to intermittent irrigation (Belder et al. 2004; Leon et al. 2021; Kato and Katsura et al. 2014). This may occur also in upland fields due to excess water after intense rainfall and cause unstable yields in crops such as soybean (Matsuo et al. 2017). SMF is perceived to be more stressful to the plant than either constant waterlogging or water deficit since the root adaptations to the predominant water stress conditions might be adversely affected by the preceding water stress conditions (Suralta et al. 2018a).

For rice, the key root traits for plant adaptation to SMF are plasticity in aerenchyma formation, root elongation, lateral root development, and root growth angle. When drought-stressed roots were subjected to sudden waterlogged conditions, the formation of aerenchyma increased, and diffusion of atmospheric O_2 to the roots was rapidly facilitated while such developmental and resulting function differed with genotypes. In addition, when previously stagnant roots were subjected to progressive soil drying conditions, root elongation and lateral root development were promoted (Suralta et al. 2008a, b, 2010, 2018a; Suralta and Yamauchi 2008; Niones et al. 2012). When the roots encountered the re-watering after drought, the genotype with high adaptation to SMF showed a rapid root elongation rate, while the ability of aerenchyma formation may be involved in this adaptation process (Kano-Nakata et al. 2013).

3.1.3 Water Deficit Due to Progress Drought

Water deficit is the major constraint for crop production (Leng and Hall 2019). The root system development plays an important role in drought avoidance (Gowda et al. 2011). Most of the research since the 1960s has identified deep rooting, thick roots, and the root with strong penetration ability (Henry 2013) as desirable root traits for drought avoidance in rice. Uga et al. (2013) cloned the *DEEPER ROOTING1* (*DRO1*) gene controlling root system architecture in rice. Introducing *DRO1* into a shallow-rooting rice cultivar increased rooting depth and yield under upland conditions.

However, in some cases, those traits may not be functional or fully expressed due to strong interaction with the soil environment under field conditions. In other words, the impact of drought patterns, particularly the timing of drought occurrence and different intensities of drought stress, may differ with fields. Thus, it is important to characterize the water environment for the identification of drought patterns. For the precise control of drought intensity, a line source sprinkler (LSS) system that can create a smooth gradient from wet to dry soil conditions is effective (Kano et al. 2011; Kameoka et al. 2016; Menge et al. 2016). Using the LSS system, we successfully evaluated root plasticity as the key root trait for adaptation to various intensities of drought stress in rice (Kano et al. 2011), and root plasticity expression was pronounced under mild drought stress conditions (Kano-Nakata et al. 2011; Menge et al. 2016).

3.1.4 Combined with Other Stresses Such as Soil Compaction and Nutrient Deficiency

Soil nutrient status substantially affects root system development and functions, and water stress and soil nutrient availability can interact with each other, which then affects the plant growth. Root plasticity expression was enhanced by N application when the drought stress was mild, but its expression was suppressed when the stress was severe (Tran et al. 2014; Menge et al. 2019). It was also enhanced under soil moisture fluctuation (Menge et al. 2020), and the enhanced root plasticity contributed to the maintenance of dry matter production under such stress conditions. Although water stress induced the variation of N forms (NH_4^+-N and NO_3^--N), root plasticity was triggered by water deficit but not by the resulting changes in the forms of soil N in rice (Tran et al. 2015). Water stress also affects the P availability in soil. Root hair plasticity triggered by P deficiency contributed to greater root growth and phosphorus acquisition in low P soils in barley (Zhu et al. 2010). Brown et al. (2012) pointed out that root hair presence was critical for sustainable yield under extreme combined P-deficit and drought stress conditions in barley.

As we mentioned before, typical rainfed lowland rice field has a hardpan, and most of the studies on root penetration ability focused on deep rooting during drought conditions (Cairns et al. 2011). Suralta et al. (2018b) clarified that root penetration into the deep soil layer due to the plasticity in nodal root elongation through the hardpan was expressed during the period when the soil moisture increased due to rewetting, resulting in the reduction in soil penetration resistance. This root plasticity contributed to increased shoot dry matter production through enhanced water uptake in deeper soil layers during the subsequent drought period (Nguyen et al. 2018).

3.1.5 Traits That Constitute Root Plasticity

The root system contains different types of roots. In case of rice, root system consists of main roots including one seminal root and nodal roots and two types of lateral roots (LRs) (L-type and S-type LRs) (Watanabe et al. 2020; Yamauchi et al. 1996). L-type LR is generally thick and long and highly branched, while S-type LR is mostly thin and short and unbranched (Kono et al. 1972). As mentioned above, in a series of our studies, we found that the root plasticity was expressed in branching and root elongation, which resulted in promoted deep rooting, aerenchyma development, and changes in rooting angles in response to various soil moisture conditions. Based on the soil water environments, we have summarized the important plastic root traits in Fig. 3.1.

Under progressive drought in upland conditions, plasticity in root elongation, which resulted in deep rooting, has been observed in rice (Kameoka et al. 2015; Menge et al. 2016, 2019; Hazman and Brown 2018), wheat (Ehdaie et al. 2012), and millet (Rostamza et al. 2013). Plasticity in lateral root development that resulted in more horizontal distribution has also been observed. In rice, plasticity in branching and promoted L type-LRs development played an important role in increasing root length density at deep soil layer (>20 cm) during progressive drought (Kameoka et al. 2015; Menge et al. 2016, 2019). In the case of the fields where the available soil was shallow (0–20 cm) and deep root growth was suppressed, plasticity in lateral root development contributed to the increased total root length. These root plasticity expressions were exhibited under mild drought stress (15–30 v/v% of soil

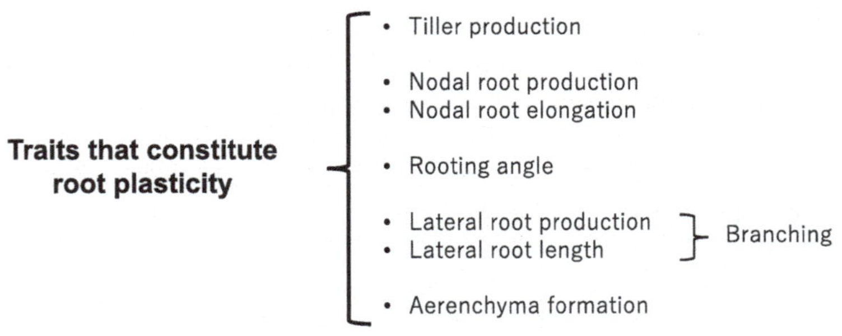

Fig. 3.1 Summary of root plasticity traits

moisture contents (SMC) (Kano et al. 2011; Tran et al. 2014, 2015) and 11–28 v/v% of SMC (Menge et al. 2016), but not under severe drought stress.

Under soil moisture fluctuations (SMF) in rainfed lowland conditions or in irrigated rice fields with water-saving cultivation technology, plasticity in nodal root production was exhibited during the time when the soil was wet due to the rainfall or irrigation (Kano-Nakata et al. 2013). This root plasticity trait was synchronized with promoted tiller production (Kano-Nakata et al. 2013). Together with nodal root production, deep rooting due to the plasticity in nodal root elongation was the key root trait under SMF when the soil moisture increased and penetration resistance decreased (Nguyen et al. 2018; Suralta et al. 2018b). Furthermore, plasticity in aerenchyma formation is one of the key traits that rapidly facilitated the diffusion of atmospheric O_2 when drought-stressed roots were subjected to sudden waterlogged conditions. Then plasticity in root elongation and lateral root development contributed to the plant water uptake when previously stagnant roots were subjected to progressive soil drying conditions (Suralta et al. 2010). Under SMF with different water distributions in the soil profile, plasticity in root growth angle was exhibited in response to the condition (Kano-Nakata et al. 2019).

These key root plasticity traits were observed to contribute to better adaptation through enhanced transpiration and photosynthesis, and consequently, dry matter production. Kameoka et al. (2016) pointed out that matching the position of root plasticity expression with soil moisture availability along the soil profile, which is dependent on the depth of soil and topography, is important for the root plasticity to be functional significant for plant growth.

3.2 Methods of Phenotyping Root Plasticity

3.2.1 Experimental System for Soil Moisture Fluctuations

The precise identification of an ideal plant phenotype depends on the preciseness of the environments being simulated. Soil moisture fluctuation (SMF) treatments for studying root morpho-physiological responses can be imposed using hydroponics, under greenhouse, and under field conditions (Table 3.1). The key is how to mimic the stresses found in the actual production growing environments such as oxygen (O_2) deficiency and/or drought stress in an experimental system simulating the actual field environments.

In hydroponics grown under either a controlled growth chamber or greenhouse, imposition of O_2 deficiency in water solution is achieved using agar (0.1% w/v) flushed with N_2 gas to prevent turbulence (convection) and simulate slow gas movement found in waterlogged soils (Wiengweera et al. 1997). On the other hand, drought is induced using polyethylene glycol 6000 (PEG 6000) (4%, W/V), which mimics the drying effects of the soil environment (Ober and Sharp 2003; Ogawa et al. 2005, Ogawa and Yamauchi 2006a, b).

Table 3.1 Experimental system for phenotyping key root traits under soil moisture fluctuations

Growing medium	Growing conditions	Soil moisture conditions	Root sampling techniques/ measurements	References
Hydroponics				
		Transient O_2 deficiency (pre-N_2 aerated and stagnant) and drought induced by PEG	Manual root extraction Manual counting of the number and measurements of lengths of the nodal roots Manual counting of L- and S-type lateral roots	Suralta et al. (2008a, b)
Soil				
Root box	Without hardpan	Well-watered- drought-rewatered	Rootbox pinboard method Manual counting of number of nodal roots Root digitization Software-aided root length measurements	Suralta et al. (2010) Kano-Nakata et al. (2012)
	With hardpan	Drought-rewatered-drought	Manual root extraction Manual counting of the number of nodal roots Root digitization Software-aided root length measurements	Nguyen et al. (2018) Suralta et al. (2018b)
Experimental bed	Flat (20 cm soil depth)	Continuous cycles of AWD through supplemental irrigation	Manual root extraction using a monolith cylinder Manual counting of the number of nodal roots Root digitization Software-aided root length measurements	Niones et al. (2012) Owusu-Nketia et al. (2018a, b)

(continued)

Table 3.1 (continued)

Growing medium	Growing conditions	Soil moisture conditions	Root sampling techniques/ measurements	References
	Sloping (32–65 cm soil depth)	Continuous cycles of AWD through supplemental irrigation	Manual root extraction using a monolith cylinder at different soil depth increments Manual counting of the number of nodal roots Root digitization Software-aided root length measurements	Owusu-Nketia et al. (2023), Kameoka et al. (2015), Menge et al. (2019)
Line source sprinkler system	Flat (20 cm soil depth)	Continuous cycles of AWD using a sprinkler irrigation system which create various intensities of soil moisture conditions perpendicular to the source of water	Manual root extraction using a monolith cylinder Manual counting of the number of nodal roots Root digitization Software-aided root length measurements	Kano-Nakata et al. (2017)
Field	Limited soil depth (20 cm)	Continuous cycles of AWD through supplemental irrigation	Manual root extraction using a monolith cylinder Manual counting of the number of nodal roots Root digitization Software-aided root length measurements	Kano-Nakata et al. (2013) Sandhu et al. (2017)
		Continuous cycles of AWD through supplemental irrigation through rainfall supply	Manual root extraction using a monolith cylinder Manual counting of the number of nodal roots Root digitization Software-aided root length measurements	Owusu-Nketia et al. (2018a, b)

In soil-filled root boxes, O_2 deficiency is achieved by immersing the rootboxes with plants in huge tanks filled with water to mimic the flooded soil environments. On the other hand, progressive drought is achieved by draining the water and allowing the soil in the rootboxes to dry down to reach certain moisture levels (Suralta et al. 2010). Using bigger soil-filled rootboxes to study, the interactions between SMF and hardpan penetration resistance on deep rooting ability (Nguyen et al. 2018; Suralta et al. 2018b), the SMF is imposed by subjecting alternately the upper soil layer to water-saturated soil conditions, then progressive drought, and then rewatered conditions for a number of days under each condition. The roots penetrating the hardpan at different periods during SMF can be monitored in real time using a pre-installed mini-rhizotron camera.

The SMF treatment simulating field condition with shallow soil layer can also be done by using either a water-tight flatbed filled with soil up to 20 cm depth (Niones et al. 2012; Owusu-Nketia et al. 2018a, b), or a sloping bed filled with soil, with no prominent groundwater table (Kameoka et al. 2015; Owusu-Nketia et al. 2023). The SMF treatment is imposed by alternate wetting and drying water management, which continues until 2 weeks after heading. The plastic sheets installed at 20 cm depth prevent the leakage of water, the deeper penetration of roots, and/or capillary rise of water from the deep to the shallow soil layer. A similar experimental setup to that of Niones et al. (2012) and Owusu-Nketia et al. (2018a, b) but with modified soil moisture treatments is also being used (Kano-Nakata et al. 2017). The installation of a sprinkler irrigation system at the center of the soil bed allows the variation in the intensities of moisture as the distance of the soil to the sprinkler increases (LSS as mentioned above). Each of the genotypes should be planted perpendicular to the sprinkler so that all plants of each genotype can be exposed to various intensities of drought during the growth period and the comparisons in growth responses can be done among different genotypes (Kano-Nakata et al. 2017).

Under actual rainfed lowland rice fields, SMF is imposed in two ways depending on the target growing conditions. The SMF can be imposed by alternate wetting and drying (Kano-Nakata et al. 2013; Niones et al. 2012; Sandhu et al. 2017; Owusu-Nketia et al. 2018a, b) with a set soil water potential during soil drying before re-irrigating the field back to well-saturated soil condition. On the other hand, alternate wetting and drying can be also achieved naturally by irrigating the field after establishment and then shifting the water supply purely from the rainfall (Owusu-Nketia et al. 2018a, b).

3.3 Quantification of Root Plasticity

Root plasticity is defined as the ability of the root system of a genotype to change its phenotypes as environmental conditions change (O'Toole and Bland 1987) resulting in promoted plant growth by mitigating the impact of stress (Yamauchi et al. 1996; Wang and Yamauchi 2006). Root plasticity can be quantified using different approaches. The reaction norm (pattern of phenotypic traits expression of a single genotype across a range of environments) (Mu et al. 2022) and the coefficient of variation (CV; standard deviation/mean) (Pélabon et al. 2020) are the most common quantification methods of plasticity. Valladares et al. (2006) reviewed the quantitative estimation of plasticity using 17 different indices such as plasticity index

(PI) and relative distances plasticity index (RDPI) and demonstrated that plasticity rankings differed when using different estimators.

As we mentioned earlier, root plasticity has been observed for different root anatomical, morphological, and developmental traits. In the study of Drew (1975), barley roots showed proliferation in response to patches of phosphate, ammonium, and nitrate. This is the study of within-plant root plasticity. In contrast, most phenotypic plasticity studies have compared multiple plants of a genotype grown under different environments. The analysis of root plasticity under water stresses can be done by comparing the roots of one genotype between stressed and non-stressed (control) conditions. Furthermore, the use of genetically similar lines such as recombinant inbred lines (RILs) (Schneider et al. 2020), chromosome segment substitution lines (CSSLs) (Owusu-Nketia et al. 2018a, b; Nguyen et al. 2018; Niones et al. 2019), and doubled haploid lines (DHLs) (Suralta et al. 2015) has resulted in precise quantitative evaluation and measurement of root plasticity with minimal effects of genetic confounding, and most of our studies cited in this chapter applied this approach.

Root plasticity based on the total root length was calculated directly as the difference between CSSL and their recurrent parent under stressed conditions in some of our studies (Kano et al. 2011; Tran et al. 2014; Kano-Nakata et al. 2017). By using CSSLs plants grown in a line source sprinkler system, we successfully quantified root plasticity under a contiguous water gradient. Our studies demonstrated that root plasticity expression was affected by the intensities of drought stress and genotypes, and root plasticity maximized at around soil water potential of −0.04 MPa for rice plants. In the study of Sandhu et al. (2016), plasticity was calculated using single replicates from the stress treatment and the mean values from the control treatment:

$$\text{Root plasticity} = \left[\frac{x_{\text{stress}} - \overline{x}_{\text{control}}}{\overline{x}_{\text{control}}} \right]$$

This approach allowed for the standardization of the data and the use of all single replicates for statistical analysis, and we applied similar approach in our studies (Nguyen et al. 2018; Suralta et al. 2018b).

3.4 Physiological Mechanism and Functional Roles of Root Plasticity

Osmotic adjustment or the accumulation of osmolyte compounds in response to water stress may be an important physiological base for root plasticity. Sugars were found to contribute to root system formation as energy source, a compatible solute, and a substance for cell wall synthesis (Ogawa et al. 2007, 2009). The accumulation of glucose and sucrose along the root apical portions, where the cell elongation zone is located, promoted lateral root development under osmotic stress conditions

(Ogawa et al. 2005), which suggested that sugar accumulation is closely associated with the root plasticity expression under water stress conditions. In the study on a rice mutant, *promoted lateral root 1* (*plr1*) (Lucob-Agustin et al. 2021), which exhibited increased lateral root development, clearly showed that reduced basal shoot starch accumulation had positive effects on lateral root development, especially L-type lateral root promotion. This is due to the repressed expression of starch biosynthesis genes and subsequent increased transport of sugars from shoots to roots, which contributed to root plasticity expression. Such carbohydrate allocation from shoot to root was observed under water deficit conditions in rice from the seedling stage (Xu et al. 2015) to maturity (Kikuta et al. 2019).

Carbohydrates used in root respiration provide the energy required for maintenance of metabolic activities for plant tissues. Araki et al. (2012) reported that root respiration increased in response to hypoxic treatments in wheat. Reduced root respiratory rates under water stress could better control the total carbon consumption. The formation of aerenchyma reduces the respiratory costs and nutrient requirements of the root utilizing the released resources for the root growth under drought or nutrient deficiency (Lynch 2015; Schneider et al. 2017a, b). Further studies are required to understand the metabolic cost as sugar consumption for root respiration and how much the promoted root development contributes to the dry matter production and crop yield under water stress conditions. Through such quantitative analysis, we can confirm how root plasticity is beneficial for plant adaptation to water stress.

3.5 Genetic Regulation of Root Plasticity

3.5.1 QTLs Controlling the Expression of Root Plasticity Interacted with Growing Environment and Growth Stages

Progress in discovery of genes or quantitative trait loci (QTLs) and the analysis of the molecular mechanism behind the root response to different stresses provide significant insights on genetic control of root development, which enhances plant adaptation to water stress and low-input agroecosystem (Price et al. 2002a, b; Wang et al. 2005; Courtois et al. 2009, 2013; Bray and Topp 2018; Guo et al. 2018; Rebouillat et al. 2009; Courdert et al. 2010; Kawai et al. 2022c).

In response to water stress, fine mappings were done for the following root QTLs (Fig. 3.2): *DRO2* and DRO*3* associated with root depth (Uga et al. 2015); *qRL6.1* for root length (Obara et al. 2010); and *qPef9* for phosphorus deficiency (Ranaivo et al. 2022); *qLLRN-12* for L-type lateral root (LR) production (Niones et al. 2019, Niones et al. 2015), *qAER-12* for aerenchyma development (Duyen et al. 2022; Niones et al. 2013), and a candidate gene on chromosome 2 for lateral root development (Niones et al. 2021) were characterized. Furthermore, *DRO1* for deep rooting

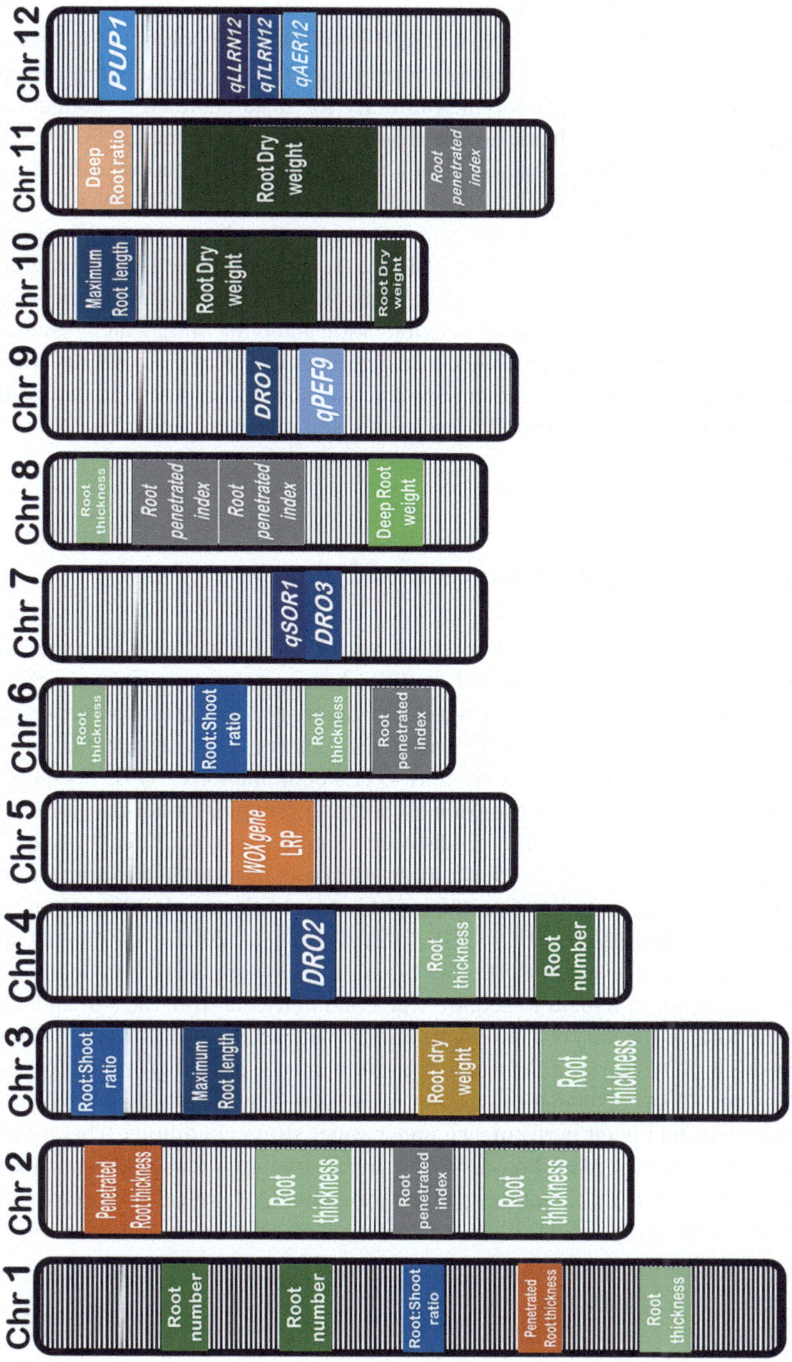

Fig. 3.2 Distribution of some quantitative trait loci (QTLs) across chromosomes associated with root traits (Duyen et al. 2022; Gamuyao et al. 2012; Hanzawa et al. 2013; Niones et al. 2013, 2015, 2019, 2021; Obara et al. 2010; Ranaivo et al. 2022; Uga et al. 2013, 2015)

(Uga et al. 2013, 2015), *PUP1* for phosphorus uptake (Gamuyao et al. 2012), and *SOR1* a gene for soil-surface rooting (Hanzawa et al. 2013) were cloned.

Kawai et al. (2022c) reported two WUSCHEL-related homeobox (*WOX*) transcription factors that opposingly regulate LR primordium size. These findings form the basis for improving root phenotypic plasticity for sustainable crop production under variable environments. In the aspect of heterorhizy (Watanabe et al. 2020), we showed L-type LR plasticity is one of the key traits for plant adaptation under fluctuating soil moisture (Niones et al. 2012; Suralta et al. 2010). The *qTLRN*-12 at the seedling stage and *qLLRN*-12 at the vegetative stage in the chromosome 12 region were closely linked with plasticity in total LR number and in L-type LR production. In addition, *qAER*-12 for aerenchyma development was found to be significantly associated with the increase in lateral root elongation and branching, which resulted in greater root system development and consequently contributed to higher dry matter production. This *qAER*-12 is presumably the first reported QTL associated with aerenchyma development in rice under fluctuating soil moisture conditions (Niones et al. 2015).

3.6 QTL Hotspots in Chromosome 12

Many QTLs associated with root morphological traits associated with tolerance to water stress in rice have been detected in several mapping populations. Courtois et al. (2009) reviewed and compiled a significant number of root QTLs covering the whole genome, which was subjected to a meta-QTL analysis, and revealed the existence of "hot spots" comprising six or more QTLs combinations. "QTL hotspots" are regions of the genome and chromosomes, wherein groups of traits are co-localized and overlapping. Chromosome 12 has been identified as one of the "hotspots" for the location of genes and QTLs that play a role in root traits related to adaptation to water stress (Courtois et al. 2003, 2009, 2013; MacMillan et al. 2006; Nguyen et al. 2004; Niones et al. 2013, 2015, 2019, 2021; Price et al. 2002a, b; Zhang et al. 2001). Root traits such as root length, root biomass, and root penetration depth are crucial for the ability of a plant to acquire water and nutrients from the soil and are therefore important for survival in environments with excess- and limited-water availability (Fig. 3.3). The significant concentration of root-related QTLs and genes on chromosome 12 suggests that this chromosome may play a central role in determining root traits and plant performance under water stress conditions. This information can be useful for plant breeders who are developing crops with improved water stress adaptation. By targeting specific genes and QTLs on chromosome 12, breeders can create crops with enhanced root systems that are better adapted to water stress environments.

Fig. 3.3 Quantitative trait loci (QTLs) detected on chromosome 12 in several mapping populations. (**a**) Courtois et al. (2003); (**b**) MacMillan et al. (2006); (**c**) Nguyen et al. (2004); (**d**) Niones et al. (2013); (**e**) Niones et al. (2015); (**f**) Price et al. (2002a) and (**g**) Zhang et al. (2001) associated with root response to water stress

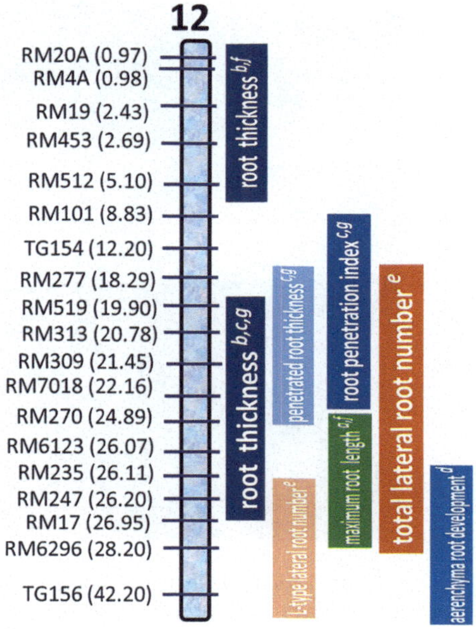

3.7 Pyramiding for Submergence and Salinity Tolerance

Multiple environmental issues limit rice productivity. In coastal areas in Southeast Asia, rainfed paddy fields are prone to water deficits during the dry season, which increases the risk of salinity stress by increasing salt concentration in the soil. During the rainy season, heavy rains can cause flash flood/submergence together with salinity stress derived from seawater that has intruded from adjacent sea water canals (Fig. 3.4). Therefore, multiple stress-tolerant rice varieties have been developed by introgressing multiple QTLs. For example, the *Sub1* QTL for submergence tolerance and the *Saltol* QTL for salinity tolerance have been introduced into various genetic backgrounds using marker-assisted-backcrossing (MABC) (de Ocampo 2022; Nair and Shylaraj 2021; Singh et al. 2021; Muthu et al. 2020; Gregorio et al. 2013). Some salinity- and submergence-tolerant two-in-one varieties have been released for commercial cultivation such as BRRI Dhan 78 in Bangladesh and Salinas 22 (IR86385–38–1-1-B) in the Philippines (Singh et al. 2021).

It is necessary to validate the effectiveness and/or interactive effect of the combined QTLs in various genetic backgrounds. For example, when combining *Saltol* QTL with *Sub1* QTL, the positive salinity tolerance effect of *Saltol* QTL was decreased in some genetic backgrounds (de Ocampo 2022; Thein 2007). However, the *Sub1* introgression improved submergence tolerance regardless of the presence of the *Saltol* QTL. Currently, it has not been determined whether the effect was related to the combination of the *Saltol* and *Sub1* QTL's or *Saltol* itself in the target rice background genotype.

Fig. 3.4 Rainfed paddy fields (right) and an adjacent brackish water channel (left) in Kampot Province, Cambodia. A dam road avoids seawater intrusion from the channel into paddy fields; however, seasonal increases in the water levels of channels can cause the overflow of seawater into the adjacent paddy fields. Additionally, heavy rain during the wet season causes flash flood or submergence, which affects rice growth as well

3.8 Molecular Mechanism of the Expression of Plasticity and Heterorhizy

As mentioned above, plants alter their root system architecture in response to the water and nutrient distribution patterns, and its molecular mechanism through the hormonal regulation has been gradually understood in *Arabidopsis* and cereals (Lucob-Agustin et al. 2021). Lateral root formation is promoted toward water patches and suppressed in dry patches or in air (Bao et al. 2014; Orman-Ligeza et al. 2018). It is known that auxin signaling is controlled by the interactions between Aux/IAA proteins, repressors of auxin-responsive transcription, and ARF transcription factors (Liscum and Reed 2002). In *Arabidopsis* roots, *LBD16*, a target gene of the ARF7, exhibits an asymmetric expression, with higher expression on the side in contact with water, which is consistent with the lateral root formation pattern (Orosa-Puente et al. 2018). It has been revealed that SUMOylation (small ubiquitin-like modifier conjugation) of ARF7, which is necessary for its interaction with the IAA3 protein that keeps the auxin signal inactive, is specifically disrupted by unidentified factor(s) on the side in contact with water, resulting in preferential *LBD16* expression and lateral root formation on the water side (Orosa-Puente et al. 2018). On the other hand, in barley roots, genes involved in plant hormone signaling, including the ABA signaling pathway, were differentially expressed, and ABA content increased after a transient water deficit. In cereals and *Arabidopsis*, transient ABA treatment decreased the free IAA levels above the root apical meristem, thereby causing the repression of pre-branch site formation (Orman-Ligeza et al. 2018).

In addition, Kawai et al. (2022c) recently reported that the LR primordium size, an important determinant of LR type, that is, heterorhizy, was indeed under the influence of two important genes belonging to the *WOX* family, *QHB/OsWOX5* and *OsWOX10*, which were found to have opposing regulatory roles, the former functioning as a brake and the latter as an accelerator in terms of the size control in rice (Fig. 3.5). They also found that the *OsWOX10* was auxin-inducible and a potential target of an ARF transcriptional factor indicating that auxin signaling upregulation in LR primordium induces *OsWOX10* expression, increasing its size (Kawai et al. 2022a). These findings are expected to lead to the development of unprecedented crop varieties with drought stress tolerance and thus contribute to securing sustainable rice production for growing population in the world.

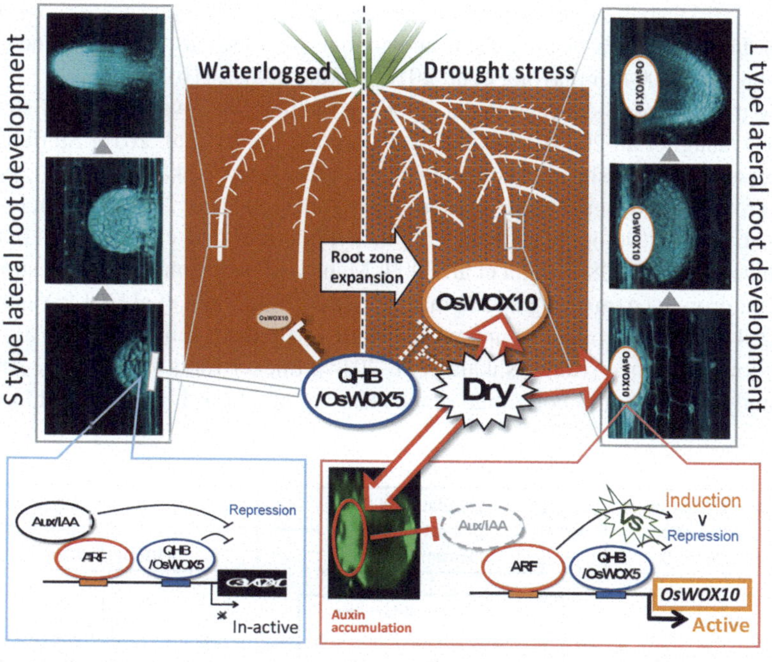

Fig. 3.5 Regulatory mechanisms for lateral root heterorhizy in rice. Under waterlogged conditions (left), *QHB/OsWOX5* represses the expression of *OsWOX10*. On the other hand, under drought stress conditions (right), auxin accumulation at the base of LR primordium induces *OsWOX10* expression, increasing its size (Based on Kawai et al. 2022a, c)

3.9 Root System Ideotype

3.9.1 Identification of Root Ideotype

An ideotype is defined as a biological model, which is expected to perform in a predictable manner within a defined environment (Donald 1968). The concept of plant type (ideotype) in fact significantly contributed to the remarkable increase of crop yield during the Green Revolution. Likewise, the concept of root ideotype was proposed by Taylor and Yamauchi (1991), Yamauchi et al. (1996), and Yamauchi (2003). Later, Gewin (2010) discussed that the root ideotypes must be custom-tailored to the growing environment. Hence, the identification of ideotype involves defining a crop production environment and designing (Gewin 2010; Taylor and Yamauchi 1991; Yamauchi 2003) a plant model combining the morphological and physiological traits that can contribute to improved plant performance and the development of an improved cultivar (Fig. 3.6). Water is essential to crop productivity. Some crop-growing environments have favorable soil moisture due to stable irrigation sources, while others experience water stress, receiving either too much (submergence or flooding) or too little water (drought) or both alternately due to erratic rainfall as source. The magnitude of the water stress also depends on the types of soil environment such as soil texture and structure (Inthavong et al. 2012; O'Geen 2013), shallow or deep soil (Kameoka et al. 2015), presence or absence of the hardpan (Hamada et al. 2021), and topography such as elevation and slope (Samson et al. 2004; Inthavong et al. 2012; Kikuta et al. 2016).

Key traits of root ideotype under flooding include the development of shallow roots (Oyanagi et al. 2004) and adventitious roots (Loreti et al. 2016; Steffens and Rasmussen 2016; Joshi and Ginzberg 2021; Mano and Nakazono 2021), aerenchyma formation (Mano and Nakazono 2021; Yamauchi et al. 2018, Yamauchi et al. 2021a, b), and development of suberin in hypodermal and exodermal cell walls (Garthwaite et al. 2008; Kotula et al. 2009; Mano and Nakazono 2021) for supporting plant growth (Armstrong and Armstrong 2014; Takahashi et al. 2014; Yamauchi et al. 2018). Under drought of upland conditions with deep soil profile, the adventitious root emergence (Kano-Nakata et al. 2014), thicker roots (Price et al. 2002a), root angle higher to the horizontal plane (Kato et al. 2006; Uga et al. 2011), deep

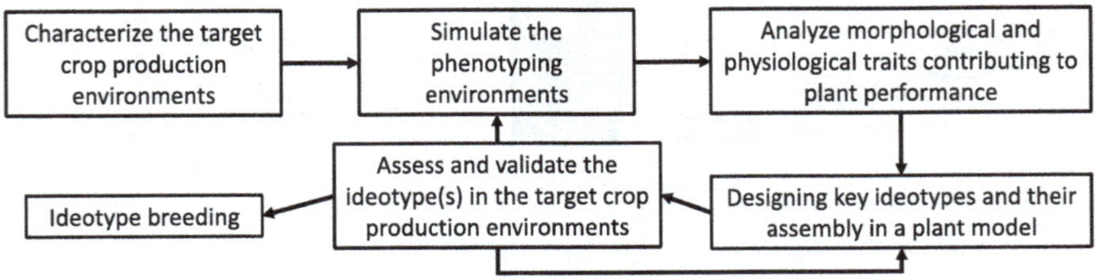

Fig. 3.6 General process flow on identifying ideotype(s) for improving plant performance in target crop production environments

and thick roots (Comas et al. 2013), deep root development (Price et al. 2002a; Kato et al. 2007, 2013; Kano-Nakata et al. 2014; Kameoka et al. 2015; Li et al. 2015), and L-type lateral root (Watanabe et al. 2020) development at depth (Kano-Nakata et al. 2014; Kameoka et al. 2015; Menge et al. 2016) are important. With limited soil depth, however, the plasticity in L-type lateral roots at the soil surface (Kano et al. 2011; Kameoka et al. 2016) is more suitable. In a rainfed lowland field without a hardpan, the plasticity in deep rooting (Henry 2013) and L-type lateral root production at deeper soil layers (Suralta et al. 2010) are more ideal. With the presence of hardpan, both the plasticity in nodal root production (Suralta et al. 2010) and lateral root production at the shallow soil layer (Kano-Nakata et al. 2013; Sandhu et al. 2017), as well as the plasticity in nodal root penetration through the hardpan (Samson et al. 2002; Suralta et al. 2018a, b; Nguyen et al. 2018) and the subsequent plasticity in L-type lateral roots below the hardpan (Suralta et al. 2018a, b; Nguyen et al. 2018), are ideal.

3.10 Breeding Approach: Introgression of Root Plasticity Traits into High-Yielding Variety to Strengthen Their Adaptability to the Stress-prone Environment

Breeding for root traits is a promising strategy to strengthen stress resilience. Selection for desirable root characteristics and exploitation of its genetic variation is of fundamental importance for breeding. Many studies revealed large genotypic variations in root development and functions that support performance under soil resources deficiency, which strongly suggests the presence of superior alleles for genes controlling root growth and nutrient uptake under such stressful conditions (Kawai et al. 2022b; Chen et al. 2020; Lynch 2019; Samejima and Tsunematsu 2016; Wissusa et al. 2016).

Some important discoveries/identification of genes/QTLs in relation to root response to different stresses include QTLs such as *DRO1*, *DRO2*, and *DRO3* associated with root depth and angle (Uga et al. 2013, 2015); *PUP1* for phosphorus uptake (Gamuyao et al. 2012); *qRL6.1* for root length (Obara et al. 2010); *qLRN12* for L-type lateral root production (Niones et al. 2012); and *qPef9* for phosphorus deficiency (Ranaivo et al. 2022). The molecular marker-assisted breeding can help shorten the timeframe for characterization and enhance the precision of identification of the number and location of QTLs (Courtois et al. 2013). Making the most of genetic diversity, marker-assisted selection, and positional QTL cloning is a useful strategy for fine-mapping and for breeding research. Root plasticity is one of the key traits for plants to adapt to the conditions of highly heterogeneous distribution in soil resources by collecting efficiently nutrients and water under such conditions (Yamauchi et al. 1996; Wang et al. 2006; Wang and Yamauchi 2006). Schneider and Lynch (2020) proposed that root phenotypic plasticity may be beneficial as a target trait for crop production in low-input systems in which the soil resources are spatio-temporally dynamic. Incorporating root traits into high-yielding mega-varieties is a

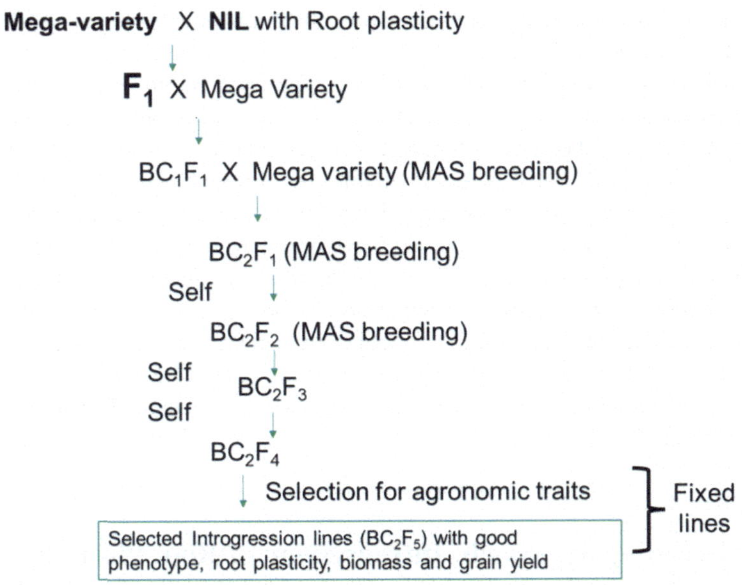

Fig. 3.7 Breeding approach on introgression of root plasticity alleles in high-yielding mega variety

promising strategy to increase crop production with strengthened climate stress resilience while enhancing resource and carbon sequestration in the soil under low-input and in water-stressed environment (Fig. 3.7).

3.11 Agronomic Approach: Fertilizer and Irrigation Management Through G × E × M Evaluation

The expression of root plasticity triggered by soil water adaptability is not only genetically regulated but also strongly regulated by genotype (G) × environment (E) interactions, which is further affected by management (M). Understanding the G × E × M interaction is essential to achieve both high stability and adaptation of rice production under spatially and temporally variable environments such as rain-fed lowland areas. Recently developed statistical analysis tools and data modeling technologies can help us predict the yield under multi-environments and close the yield gap between attainable potential yield and farmers' yield (Cooper et al. 2021). Through G × E × M analysis, accurate identification of root plasticity directly linked with yield could be possible (Wade et al. 2015; Sandhu et al. 2016).

As irrigation management, water-saving technologies such as alternative wet and dry (AWD) and aerobic culture can reduce the water used for rice cultivation but often suffer from yield penalty relative to continuously flooded conditions (Carrijo et al. 2017). The genotypes with high root plasticity should be potential materials that could reduce water use without yield loss. However, root plasticity could maintain shoot dry matter production under mild drought stress, but not under severe drought stress (Suralta et al. 2018a, b), indicating the importance of good water management to maximize the beneficial effects of root plasticity expression.

As we mentioned previously, water-by-nutrient interactions affected the expression of root plasticity, and nitrogen application enhanced root plasticity under mild drought conditions (Tran et al. 2014, 2015). However, increased fertilizer application has negative impacts on the environment. Recently, together with AWD, biological nitrification inhibition (BNI) technology and cattle biogas effluent application are getting attention as a sustainable technology that enable to reduce soil CH_4 and N_2O emissions without yield loss (Minamikawa et al. 2021; Subbarao and Searchinger 2021). Therefore, the type, level, and timing of fertilizer application should be optimized to maximize plant growth through root plasticity expression considering both yield advantages and soil chemistry.

Perdiguerra et al. (2021) reported that environmental differences had a higher contribution to the variations in rice grain yield compared to genotypic variations, and climatic factors such as solar radiation and air temperature mostly influenced rice growth and productivity under field conditions. Thus, firstly, the target stress-prone environment should be fully characterized, and then production technology that can enhance root plasticity expression and function in each target environment needs to be established.

References

Araki H, Hossain MA, Takahashi T (2012) Waterlogging and hypoxia have permanent effects on wheat root growth and respiration. J Agron Crop Sci 198:264–275

Armstrong W, Armstrong J (2014) Plant internal oxygen transport (diffusion and convection) and measuring and modelling oxygen gradients. In: van Dongen, J.T., Licausi, F. (eds) Low-oxygen stress in plants: sensing and adaptive responses to hypoxia. Plant Cell Monogr 21:267–298

Bao Y, Aggarwal P, Robbins NE, Sturrock CJ, Thompson MC, Tan HQ, Tham C, Duan L, Rodriguez PL, Vernoux T, Mooney SJ, Bennett MJ, Dinneny JR (2014) Plant roots use a patterning mechanism to position lateral root branches toward available water. Proc Natl Acad Sci U S A 111:9319–9324

Belder P, Bouman BAM, Cabangon R, Guoan L, Quilang EJP, Li Y, Spiertz JHJ, Tuong TP (2004) Effect of water-saving irrigation on rice yield and water use in typical lowland conditions in Asia. Agric Water Manag 65:193–210

Bray AL, Topp CN (2018) The quantitative genetic control of root architecture in maize. Plant Cell Physiol 59:1919–1930. https://doi.org/10.1093/pcp/pcy141

Brown LK, George TS, Thompson JA, Wright G, Lyon J, Dupuy L, Hubbard SF, White PJ (2012) What are the implications of variation in root hair length on tolerance to phosphorus deficiency in combination with water stress in barley (*Hordeum vulgare*)? Ann Bot 110:319–328

Cairns JE, Impa SM, O'Toole JC, Jagadish SVK, Price AH (2011) Influence of the soil physical environment on rice (*Oryza sativa* L.) response to drought stress and its implications for drought research. Field Crop Res 121:303–310

Carrijo DR, Lundy ME, Linquist BA (2017) Rice yields and water use under alternate wetting and drying irrigation: a meta-analysis. Field Crop Res 203:173–180

Chen J, Wang Z, Zhang Y, Sun H, Song S, Bai Z, Lu Z, Li C (2020) Nitrogen fertilization root growth and coordinates the root-shoot relations in cotton. Front Plant Sci 11:880. https://doi.org/10.3389/fpls.2020.00880

Chen L, Zhu T, Ogorek TLP, Wang Y, Sauter M, Pedersen O (2022) The pyramiding of three key traits aid breeding of flood-tolerant rice. Plants (Basel) 11(15):2033. https://doi.org/10.3390/plants11152033

Comas L, Becker SR, Cruz VMV, Byrne PF (2013) Root traits contributing to plant productivity under drought. Front Plant Sci 4:442

Cooper M, Voss-Fels KP, Messina CD, Tang T, Hammer GL (2021) Tackling G × E × M interactions to close on-farm yield-gaps: creating novel pathways for crop improvement by predicting contributions of genetics and management to crop productivity. Theor Appl Genet 134:1625–1644

Courdert Y, Perin C, Courtois B, Khong NG, Gantet P (2010) Genetic control of root development in rice, the model cereal. Trends Plant Sci 15:219–226

Courtois B, Shen L, Petalcorin W, Carandang S, Mauleon R, Li Z (2003) Locating QTLs controlling constitutive root traits in the rice population IAC 165 × Co39. Euphytica 134:335–345

Courtois B, Ahmadi N, Khowaja F, Price AH, Rami JF, Frouin J, Hamelin C, Ruiz M (2009) Rice root genetic architecture: meta-analysis from a drought QTL database. Rice 2:115–128

Courtois B, Audebert A, Dardou A, Roques S, Herrera TG, Droc G, Frouin J, Rouan L, Goze E, Kilian A, Ahmadi N, Dingkuhn M (2013) Genome-wide association mapping of root traits in a japonica rice panel. PLoS One 8(11):e78037. https://doi.org/10.1371/journal.pone.0078037

de Ocampo MP (2022) Identification of novel QTLs and introgression through marker-assisted backcrossing for improved salinity tolerance in rice. [Doctoral dissertation, Nagoya University]

Donald CM (1968) The breeding of crop ideotypes. Euphytica 17:385–403

Drew MC (1975) Comparison of the effects of a localised supply of phosphate, nitrate, ammonium and potassium on the growth of the seminal root system, and the shoot, in barely. New Phytol 75:479–490

Duyen DV, Kwon Y, Kabange NR, Lee J-Y, Lee S-M, Kang J-W, Park H, Cha J-K, Cho J-H, Shin D, Lee J-H (2022) Novel QTL associated with aerenchyma-mediated radial oxygen loss (ROL) in rice (*Oryza sativa* L.) under iron (II) sulfide. Plants 11(6):788. https://doi.org/10.3390/plants11060788

Ehdaie B, Layne AP, Waines JG (2012) Root plasticity to drought influences grain yield in wheat. Euphytica 186:219–232

Galamay TO, Yamauchi S, Nonoyama T, Kono Y (1992) Acropetal lignification in cereal nodal root axes as affected by different soil moisture conditions. Jpn J Crop Sci 61:511–517

Gamuyao R, Chin JH, Tanaka JP, Pesaresi P, Catausan S, Dalid C, Loedin IS, Mendoza EMT, Wissuwa M, Heuer S (2012) The protein kinase Pstol1 from traditional rice confers tolerance of phosphorus deficiency. Nature 88(7412):535–539. https://doi.org/10.1038/nature11346

Garthwaite AJ, Armstrong W, Colmer TD (2008) Assessment of O_2 diffusivity across the barrier to radial O_2 loss in adventitious roots of *Hordeum marinum*. New Phytol 179:405–416

Gewin V (2010) An underground revolution. Nature 466:552–553

Gowda VRP, Henry A, Yamauchi A, Shashidhar HE, Serraj R (2011) Root biology and genetic improvement for drought avoidance in rice. Field Crop Res 122:1–13

Gregorio GB, Islam MR, Vergara GV, Thirumeni S (2013) Recent advances in rice science to design salinity and other abiotic stress tolerant rice varieties. SABRAO J Breed Genet 45:31–41

Guo J, Chen L, Li YX, Shi YS, Song YC, Zhang DF, Li Y, Wang TY, Yang DG, Li CH (2018) Meta-QTL analysis and identification of candidate genes related to root traits in maize. Euphytica 214:223. https://doi.org/10.1007/s10681-018-2283-3

Hamada K, Inoue H, Miyamoto T (2021) Effect of hardpan on the vertical distribution of water stress in a converted paddy field. Soil Tillage Res 214:105161

Hanzawa E, Sasaki K, Nagai S, Obara M, Fukuta Y, Uga Y, Miyao A, Hirochika H, Higashitani A, Maekawa M, Sato T (2013) Isolation of novel mutant gene for soil-surface rooting in rice (*Oryza sativa*). Rice 6:30. https://doi.org/10.1186/1939-8433-6-30

Hayashi T (2014) Functional roles of root system in excess moisture tolerance in common wheat. Ph.D. thesis, Nagoya University

Hayashi T, Yoshida T, Fujii K, Mitsuya S, Tsuji T, Okada Y, Hayashi E, Yamauchi A (2013) Maintained root length density contributes to the waterlogging tolerance in common wheat (*Triticum aestivum* L.). Field Crop Res 152:27–35

Hazman M, Brown KM (2018) Progressive drought alters architectural and anatomical traits of rice roots. Rice 11:62

Henry A (2013) IRRI's drought stress research in rice with emphasis on roots: accomplishments over the last 50 years. Plant Roots 7:15–19

Inthavong T, Tsubo M, Fukai S (2012) Soil clay content, rainfall, and toposequence positions determining spatial variation in field water availability as estimated by a water balance model for rainfed lowland rice. Crop Past Sci 63:529–538

Joshi M, Ginzberg I (2021) Adventitious root formation in crops—potato as an example. Physiol Plant 172:124–133

Kameoka E, Suralta RR, Mitsuya S, Yamauchi A (2015) Matching the expression of root plasticity with soil moisture availability maximizes production of rice plants grown in an experimental sloping bed with soil moisture gradients. Plant Prod Sci 18:267–276

Kameoka E, Suralta RR, Mitsuya S, Yamauchi A (2016) Developmental plasticity of rice root system grown under mild drought stress condition with shallow soil depth; comparison between nodal and lateral roots. Plant Prod Sci 19:411–419

Kano M, Inukai Y, Kitano H, Yamauchi A (2011) Root plasticity as the key root trait for adaptation to various intensities of drought stress in rice. Plant Soil 34:117–128

Kano-Nakata M, Inukai Y, Wade LJ, Siopongco JDLC, Yamauchi A (2011) Root development, water uptake, and shoot dry matter production under water deficit conditions in two CSSLs of rice: functional roles of root plasticity. Plant Prod Sci 14:307–317

Kano-Nakata M, Suralta RR, Niones JM, Yamauchi A (2012) Root sampling by using a root box-pinboard method. In: Shashidhar HE, Henry A, Hardy B (eds) Methodologies for root drought studies in rice. International Rice Research Institute, Los Banos, Philippines, pp 34–39

Kano-Nakata M, Gowda VRP, Henry A, Serraj R, Inukai A, Suralta RR, Yamauchi A (2013) Functional roles of the plasticity of root system development in biomass production and water uptake under rainfed lowland conditions. Field Crop Res 144:288–296

Kano-Nakata M, Tatsumi J, Inukai Y, Asanuma S, Yamauchi A (2014) Effect of various intensities of drought stress on δ13C variation among plant organs in rice. Comparison of two cultivars. 2014. Am J Plant Sci 5:1686–1693

Kano-Nakata M, Inukai Y, Siopongco JDLC, Mitsuya S, Yamauchi A (2017) Quantitative evaluation of plastic root responses to contiguous water gradient in rice. Plant Root 11:70–78

Kano-Nakata M, Nakamura T, Mitsuya S, Yamauchi A (2019) Plasticity in root system architecture of rice genotypes. Exhibited under different soil water distributions in soil profile. Plant Prod Sci 22:501–509

Kato Y, Abe J, Kamoshita A, Yamagishi J (2006) Genotypic variation in root growth angle in rice (*Oryza sativa* L.) and its association with deep root development in upland fields with different water regimes. Plant Soil 287:117–129

Kato Y, Kamoshita A, Yamagishi J, Imoto H, Abe J (2007) Growth of rice (*Oryza sativa* L.) cultivars under upland conditions with different levels of water supply 3. Root system development, soil moisture change and plant water status. Plant Prod Sci 10:3–13

Kato Y, Katsura K (2014) Rice adaptation to aerobic soils: Physiological considerations and implications for agronomy. Plant Prod Sci 17:1–12

Kato Y, Tajima R, Homma K, Toriumi A, Yamagishi J, Shiraiwa T, Mekwatanakarn P, Jongdee B (2013) Root growth response of rainfed lowland rice to aerobic conditions in northeastern Thailand. Plant Soil 368:557–567

Kawai T, Akahoshi R, Shelley IJ, Kojima T, Sato M, Tsuji H, Inukai Y (2022a) Auxin distribution in lateral root primordium development affects its size and lateral root diameter in rice. Front Plant Sci 13:834378

Kawai T, Chen Y, Takahashi H, Inukai Y, Siddique KHM (2022b) Rice genotypes express compensatory root growth with altered root distribution in response to root cutting. Front Plant Sci 13:830577. https://doi.org/10.3389/fpls.2022.830577

Kawai T, Shibata K, Akahoshi R, Nishiuchi S, Takahashi H, Nakazono M, Kojima T, Nosaka-Takahashi M, Sato Y, Toyoda A, Lucob-Agustin N, Kano-Nakata M, Suralta RR, Niones JM, Chen Y, Siddique KHM, Yamauchi A, Inukai Y (2022c) *WUSCHEL*-related homeobox family genes in rice control lateral root primordium size. Proc Natl Acad Sci U S A 119:e2101846119

Kikuta M, Makihara D, Arita N, Miyazaki A, Yamamoto A (2019) Effects of different water management practices on the dry matter production process and characteristics in NERICAs. Plant Prod Sci 22:168–179

Kikuta M, Yamamoto Y, Pasolon YB, Rembon FS, Miyazaki A, Makihara D (2016) How growth and yield of upland rice vary with topographic conditions. A case of slash-and-burn rice farming in South Konawe Regency, Southeast Sulawesi Province, Indonesia. Trop Agric Dev 60:162–171

Kono Y, Igeta M, Yamada N (1972) Studies on the developmental physiology of the lateral roots in rice seminal roots. Jpn J Crop Sci 41:192–204

Kono Y, Yamauchi A, Kawamura N, Tatsumi J, Nonoyama T, Inagaki N (1987) Interspecific differences of the capacities of waterlogging and drought tolerance among summer cereals. Jpn J Crop Sci 56:115–129

Kono Y, Yamauchi A, Nonoyama T, Tatsumi J (1988) Comparison of growth responses to waterlogging of summer cereals with special reference to rooting ability. Jpn J Crop Sci 57:321–331

Kotula L, Ranathunge K, Steudle E (2009) Apoplastic barriers effectively block oxygen permeability across outer cell layers of rice roots under deoxygenated conditions: roles of apoplastic pores and of respiration. New Phytol 184:909–917

Leng G, Hall J (2019) Crop yield sensitivity of global major agricultural countries to droughts and the projected changes in the future. Sci Total Environ 654:811–821

Leon A, Minamikawa K, Izumi T, Chiem NH (2021) Estimating impacts of alternate wetting and drying on greenhouse gas emissions from early wet rice production in a full-dike system in an Giang Province, Vietnam, through life cycle assessment. J Clean Prod 285:125309

Li J, Han Y, Liu L, Chen Y, Du Y, Zhang J, Sun H, Zhao Q (2015) *qRT9*, a quantitative trait locus controlling root thickness and root length in upland rice. J Exp Bot 66:2723–2732

Liscum E, Reed J (2002) Genetics of AUX/IAA and ARF action in plant growth and development. Plant Mol Biol 49:387–400

Loreti E, van Veen H, Perata P (2016) Plant responses to flooding stress. Curr Opin Plant Biol 33:64–71

Lucob-Agustin N, Kawai T, Kano-Nakata M, Suralta RR, Niones JM, Hasegawa T, Inari-Ikeda M, Yamauchi A, Inukai Y (2021) Morpho-physiological and molecular mechanisms of phenotypic root plasticity for rice adaptation to water stress conditions. Breed Sci 71:20–29

Lynch JP (2015) Root phenes that reduce the metabolic costs of soil ex- ploration: opportunities for 21st century agriculture. Plant Cell Environ 38:1775–1784

Lynch JP (2019) Root phenotypes for improved nutrient capture: an underexploited opportunity for global agriculture. New Phytol 223:548–564

MacMillan K, Emrich K, Piepho HP, Mullins CE, Price AH (2006) Assessing the importance of genotype × environmental interaction for root traits in rice using a mapping population II: conventional QTL analysis. Theor Appl Genet 113:953–964

Mano Y, Nakazono M (2021) Genetic regulation of root traits for soil flooding tolerance in genus *Zea*. Breed Sci 71:30–39

Matsuo N, Takahashi M, Yamada T, Takahashi M, Hajika M, Fukami K, Tsuchiya S (2017) Effects of water table management and row width on the growth and yield of three soybean cultivars in southwestern Japan. Agric Water Manag 192:85–97

Menge DM, Kameoka E, Kano-Nakata M, Yamauchi A, Asanuma S, Asai H, Kikuta M, Suralta RR, Koyama T, Tran TT, Siopongco JDLC, Mitsuya S, Inukai Y, Makihara D (2016) Drought-induced root plasticity of two upland NERICA varieties under conditions with contrasting soil depth characteristics. Plant Prod Sci 19:389–400

Menge DM, Onyango JC, Yamauchi A, Kano-Nakata M, Asanuma S, Tran TT, Inukai Y, Kikuta M, Makihara D (2019) Effect of nitrogen application on the expression of drought induced root plasticity of upland NERICA rice. Plant Prod Sci 22:180–191

Menge DM, Kano-Nakata M, Yamauchi A, Suralta RR, Makihara D (2020) Root and shoot responses of upland new rice for Africa varieties to fluctuating soil moisture conditions as affected by different levels of nitrogen fertilization. J Agron Crop Sci 206:322–337

Minamikawa K, Huynh KC, Uno K, Tran NS, Nguyen CH (2021) Cattle biogas effluent application with multiple drainage mitigates methane and nitrous oxide emissions from a lowland rice paddy in the Mekong Delta, Vietnam. Agric Ecosyst Environ 319:107568

Mu Q, Guo T, Li X, Yu J (2022) Phenotypic plasticity in plant height shaped by interaction between genetic loci and diurnal temperature range. New Phytol 233:1768–1779

Muthu V, Abbai R, Nallathambi J, Rahman H, Ramasamy S, Kambale R, Thulasinathan T, Ayyenar B, Muthurajan R (2020) Pyramiding QTLs controlling tolerance against drought, salinity, and submergence in rice through marker assisted breeding. PLoS One 15(1):e0227421

Nair MM, Shylaraj KS (2021) Introgression of dual abiotic stress tolerance QTLs (Saltol QTL and Sub1 gene) into rice (*Oryza sativa* L.) variety Aiswarya through marker assisted backcross breeding. Physiol Mol Biol Plants 27:497–514

Nguyen TTT, Klueva N, Chamareck V, Aarti A, Magpantay G, Millena ACM, Pathan MS, Nguyen HT (2004) Saturation mapping of QTL regions and identification of putative candidate genes for drought tolerance. Mol Gen Genomics 272:35–46

Nguyen DT, Suralta RR, Kano-Nakata M, Mitsuya S, Owusu-Nketia S, Yamauchi A (2018) Genotypic variations in the plasticity of nodal root penetration through the hardpan during soil moisture fluctuations among four rice varieties. Plant Prod Sci 21:93–105

Niones JM, Suralta RR, Inukai Y, Yamauchi A (2012) Field evaluation on functional roles of root plastic responses on dry matter production and grain yield of rice under cycles of transient soil moisture stresses using chromosome segment substitution lines. Plant Soil 359:107–120

Niones JM, Inukai Y, Suralta RR, Yamauchi A (2013) Roles of root aerenchyma development and its associated QTL in dry matter production under transient moisture stress in rice. Plant Prod Sci 16:205–216

Niones JM, Inukai Y, Suralta RR, Yamauchi A (2015) QTL associated with lateral root plasticity in response to soil moisture fluctuation stress in rice. Plant Soil 391:63–75

Niones JM, Suralta RR, Inukai Y, Kano-Nakata M, Yamauchi A (2019) 'Kasalath' allele in Nipponbare background is responsible for the plasticity in lateral root development of rice under soil moisture fluctuation stress. Philipp Agric Sci 102:188–198

Niones JM, Lipio PLG, Cruz AS, Cabral MCJ, Hautea DM, Agustin NL, Suralta RR (2021) Genome-wide association mapping for the identification of SNPs controlling lateral root plasticity in selected rice germplasms of The Philippines. Philipp J Sci 150:663–674

O'Geen AT (2013) Soil water dynamics. Nat Educ Knowl 4:9

O'Toole JC, Bland WL (1987) Genotypic variation in crop plant root systems. Adv Agron 41:91–145

Obara M, Tamura W, Ebitani T, Yano M, Sato T, Yamaya T (2010) Fine-mapping of *qRL6.1*, a major QTL for root length of rice seedlings grown under a wide range of NH4(+) concentrations in hydroponic conditions. Theor Appl Genet 121:535–547. https://doi.org/10.1007/s00122-010-1328-3

Ober ES, Sharp RE (2003) Electrophysiological responses of maize roots to low water potentials: relationship to growth and ABA accumulation. J Exp Bot 54:813–824

Ogawa A, Yamauchi A (2006a) Root osmotic adjustment under osmotic stress in maize seedlings. 1. Transient change of growth and water relations in roots in response to osmotic stress. Plant Prod Sci 9:27–38

Ogawa A, Yamauchi A (2006b) Root osmotic adjustment under osmotic stress in maize seedlings. 2. Mode of accumulation of several solutes for osmotic adjustment in the root. Plant Prod Sci 9:39–46

Ogawa A, Kawashima C, Yamauchi A (2005) Sugar accumulations along the seminal root axis, as affected by osmotic stress in maize: A possible physiological basis for plastic lateral root development. Plant Prod Sci 8:73–180

Ogawa A, Kawashima C, Kinji K, Toyofuku K, Yamauchi A (2007) The role of osmotic adjustment, cell production, and sugar accumulation for the root growth under the osmotic stress condition. Plant Root 16:47–58. (In Japanese with English abstract)

Ogawa A, Ando F, Toyofuku K, Kawashima C (2009) Sucrose metabolism for the development of seminal root in maize seedlings. Plant Prod Sci 12:9–16

Orman-Ligeza B, Morris EC, Parizot B, Lavigne T, Babé A, Ligeza A, Klein S, Sturrock C, Xuan W, Novák O, Ljung K, Fernandez MA, Rodriguez PL, Dodd IC, De Smet I, Chaumont F, Batoko H, Périlleux C, Lynch JP, Bennett MJ, Beeckman T, Draye X (2018) The xerobranching response represses lateral root formation when roots are not in contact with water. Curr Biol 28:3165–3173.e5

Orosa-Puente B, Leftley N, von Wangenheim N, Banda J, Srivastava AK, Hill K, Truskina J, Bhosale R, Morris E, Srivastava M, Kümpers B, Goh T, Fukaki H, Vermeer JEM, Vernoux T, Dinneny JR, French AP, Bishopp A, Sadanandom A, Bennett MJ (2018) Root branching toward water involves posttranslational modification of transcription factor ARF7. Science 362:1407–1410

Owusu-Nketia S, Siangliw JL, Siangliw M, Toojinda T, Vanavichit A, Ratsameejanphen N, Ruangsiri M, Sriwiset S, Suralta RR, Inukai Y, Mitsuya S, Kano-Nakata M, Nguyen DTN, Takuya K, Yamauchi A (2018a) Functional roles of root plasticity and its contribution to water uptake and dry matter production of CSSLs with the genetic background of KDML105 under soil moisture fluctuation. Plant Prod Sci 21:266–277

Owusu-Nketia S, Inukai Y, Suralta RR, Doi K, Mitsuya S, Kano-Nakata M, Niones JM, Nguyen DTN, Kabuki T, Makihara D, Yamauchi A (2018b) Root plasticity under fluctuating soil moisture stress exhibited by backcross inbred line of a rice variety, Nipponbare carrying introgressed segments from KDML105 and detection of the associated QTLs. Plant Prod Sci 21:106–122

Owusu-Nketia S, Suralta RR, Inukai Y, Mitsuya S, Kano-Nakata M, Ofori PA, Nguyen DTN, Akira Yamauchi A (2023) Dry matter production of two rice cultivars with contrasting root plasticity expression under different topographic conditions subjected to soil moisture fluctuation. Philipp Agric Scientist 106:131–142

Oyanagi A, Kiribuchi-Otobe C, Yanagisawa T, Miura S, Kobayashi H, Muranaka S (2004) Growth and grain yield of wheat experimental lines with deep and shallow root system in wet paddy fields. Jpn J Crop Sci 73:300–308. *(In Japanese with English abstract)*

Pedersen O, Sauter M, Colmer TD, Nakazono M (2021) Regulation of root adaptive anatomical and morphological traits during low soil oxygen. New Phytol 229:42–49. https://doi.org/10.1111/nph.16375

Pélabon C, Hilde CH, Einum S, Gamelon M (2020) On the use of the coefficient of variation to quantify and compare trait variation. Evol Lett 4:180–188

Perdiguerra KNC, Mitsuya S, Yamauchi A, Lansigan FLP, Espaldon MVO, Hernandez JE, Sta. Cruz PC (2021) Grain yield variations in rice genotypes under different growing environments in The Philippines. Philipp Agric Sci 104:135–145

Price AH, Steele KA, Gorham J, Bridges JM, Moore BJ, Evans JL, Richardson P, Jones RGW (2002a) Upland rice grown in soil-filled chambers and exposed to contrasting water-deficit regimes: I. Root distribution, water use and plant water status. Field Crop Res 76:11–24

Price AH, Steele KA, Moore BJ, Jones RGW (2002b) Upland rice grown in soil filled chambers and exposed to contrasting water-deficit regimes II. Mapping quantitative trait loci for root morphology and distribution. Field Crop Res 76:25–43

Ranaivo HN, Lam DT, Ueda Y, Tanaka JP, Takanashi H, Ramanankaja L, Razafimbelo T, Wissuwa M (2022) QTL mapping for early root and shoot vigor of upland rice (*Oryza sativa* L.) under

P deficient field condition in Japan and Madagascar. Front Plant Sci 13:1017419. https://doi.org/10.3389/fpls.2022.1017419

Rebouillat J, Dievart A, Verdeil JL, Escoute J, Giese G, Breitler JC, Gantet P, Espeout S, Guiderdoni E, Périn C (2009) Molecular genetics of rice root development. Rice 2:15–34

Rostamza M, Richards RA, Watt M (2013) Response of millet and sorghum to a varying water supply around the primary and nodal roots. Ann Bot 112:439–446

Samejima H, Tsunematsu H (2016) Genotypic variation in rice varieties screened for deep rooting under field conditions in West Africa. Plant Prod Sci 19:181–192

Samson BK, Hasan M, Wade LJ (2002) Penetration of hardpans by rice lines in the rainfed lowlands. Field Crop Res 76:175–188

Samson BK, Ali A, Rashid MA, Mazid MA, Wade LJ (2004) Topographic position influences water availability in rainfed lowland rice in Rajshahi, Northwest Bangladesh. Plant Prod Sci 7:101–103

Sandhu N, Raman KA, Torres RO, Audebert A, Dardou A, Kumar A, Henry A (2016) Rice root architectural plasticity traits and genetic regions for adaptability to variable cultivation and stress conditions. Plant Physiol 171:2562–2576

Sandhu N, Subedi SR, Yadaw RB, Chaudhary B, Prasai H, Iftekharuddaula K, Thanak T, Thun V, Battan KR, Ram M, Venkateshwarlu C, Lopena V, Pablico P, Maturan PC, Cruz MTS, Raman KA, Collard B, Kumar A (2017) Root traits enhancing rice grain yield under alternate wetting and drying condition. Front Plant Sci 8:1879

Schneider HM, Postma JA, Wojciechowski T, Kuppe C, Lynch JP (2017a) Root cortical senescence improves growth under suboptimal availability of N, P, and K. Plant Physiol 174:2333–2347

Schneider HM, Wojciechowski T, Postma JA, Brown KM, Lücke A, Zeisler V, Schreiber L, Lynch JP (2017b) Root cortical senescence de- creases root respiration, nutrient content and radial water and nutrient transport in barley. Plant Cell Environ 40:1392–1408

Schneider HM, Lynch JP (2020) Should root plasticity be a crop breeding target? Front Plant Sci 11:546. https://doi.org/10.3389/fpls.2020.00546

Schneider HM, Klein SP, Hanlon MT, Nord EA, Kaeppler S, Brown KM, Warry A, Bhosale R, Lynch JP (2020) Genetic control of root architectural plasticity in maize. J Exp Bot 71:3185–3197

Singh RK, Kota S, Flowers TJ (2021) Salt tolerance in rice: seedling and reproductive stage QTL mapping come of age. Theor Appl Genet 134:3495–3533

Steffens B, Rasmussen A (2016) The physiology of adventitious roots. Plant Physiol 170:603–617

Subbarao GV, Searchinger TD (2021) A "more ammonium solution" to mitigate nitrogen pollution and boost crop yields. Proc Natl Acad Sci U S A 118:e2107576118

Suralta RR, Yamauchi A (2008) Root growth, aerenchyma development, and oxygen transport in rice genotypes subjected to drought and waterlogging. Environ Exp Bot 64:75–82

Suralta RR, Inukai Y, Yamauchi A (2008a) Genotypic variations in responses of lateral root development to transient moisture stresses in rice cultivars. Plant Prod Sci 11:324–335

Suralta RR, Inukai Y, Yamauchi A (2008b) Utilizing chromosome segment substitution lines (CSSLs) for evaluation of root responses to transient moisture stresses in rice. Plant Prod Sci 11:457–465

Suralta RR, Inukai Y, Yamauchi A (2010) Dry matter production in relation to root plastic development, oxygen transport and water uptake of rice under transient soil moisture stresses. Plant Soil 332:87–104

Suralta RR, Lucob NB, Niones JM, Perez LM, Nguyen H (2015) Developmental and quantitative trait loci analyses of root plasticity in response to soil moisture fluctuation in rice. Philippine J Crop Sci 40:12–24

Suralta RR, Kano-Nakata M, Niones JM, Inukai Y, Kameoka E, Tran TT, Menge M, Mitsuya S, Yamauchi A (2018a) Root plasticity for maintenance of productivity under abiotic stressed soil environments in rice: Progress and prospects. Field Crop Res 220:57–66. https://doi.org/10.1016/j.fcr.2016.06.023

Suralta RR, Niones JM, Kano-Nakata M, Tran TT, Mitsuya S, Yamauchi A (2018b) Root plasticity in nodal root elongation through the hardpan triggered by rewatering during soil moisture fluctuation stress in rice. Nat Sci Rep 8:4341

Takahashi H, Yamauchi T, Colmer TD, Nakazono M (2014) Aerenchyma formation in plants. In: van Dongen JT, Licausi F (eds) Low-oxygen stress in plants: sensing and adaptive responses to hypoxia, vol 21, (Plant Cell Monographs). Springer, pp 247–265. https://doi.org/10.1007/978-3-7091-1254-0_13

Taylor HM, Yamauchi A (1991) Opportunities and problems in manipulating cereal root systems. In: Acevedo et al (eds) Physiology-breeding of winter cereals for stressed Mediterranean environments. INRA, Paris, pp 369–403

Thein SZM (2007) Introgression of salinity (Saltol) and submergence (Sub1) tolerance to new genetic background of rice (*Oryza sativa* L.). [Master's thesis, University of Philippines, Los Baños]

Tran TT, Kano-Nakata M, Takeda M, Menge D, Mitsuya S, Inukai Y, Yamauchi A (2014) Nitrogen application enhanced the expression of developmental plasticity of root systems triggered by mild drought stress in rice. Plant Soil 378:139–152

Tran TT, Kano-Nakata M, Suralta RR, Menge D, Mitsuya S, Inukai Y, Yamauchi A (2015) Root plasticity and its functional roles were triggered by water deficit but not by the resulting changes in the forms of soil N in rice. Plant Soil 386:65–76

Uga Y, Okuno K, Yano M (2011) Dro1, a major QTL involved in deep rooting of rice under upland conditions. J Exp Bot 62:2485–2894

Uga Y, Sugimoto K, Ogawa S, Rane J, Ishitani M, Hara N, Kitomi Y, Inukai Y, Ono K, Kanno N, Inoue H, Takehisa H, Motoyama R, Nagamura Y, Wu J, Matsumoto T, Takai T, Okuno K, Yano M (2013) Control of root system architecture by DEEPER ROOTING 1 increases rice yield under drought conditions. Nat Genet 45:1097–1102

Uga Y, Kitomi Y, Yamamoto E, Kanno N, Kawai S, Mizubayashi T, Fukuoka S (2015) A QTL for root growth angle on rice chromosome 7 is involved in the genetic pathway of DEEPER ROOTING 1. Rice 8:8. https://doi.org/10.1186/s12284-015-0044-7

Valladares F, Sanchez-Gomez D, Zavala MA (2006) Quantitative estimation of phenotypic plasticity: bridging the gap between the evolutionary concept and its ecological applications. J Ecol 94:1103–1116

Wade LJ, George T, Ladha JK, Singh U, Bhuiyan SI, Pandey S (1998) Opportunities to manipulate nutrient by water interactions in rainfed lowland rice systems. Field Crop Res 56:93–112

Wade LJ, Bartolome V, Mauleon R, Vasant VD, Prabakar SM, Chelliah M, Kameoka E, Nagendra K, Reddy KRK, Varma CMK, Patil KG, Shrestha R, Al-Shugeairy Z, Al-Ogaidi F, Munasinghe M, Gowda V, Semon M, Suralta RR, Shenoy V, Vadez V, Serraj R, Shashidhar HE, Yamauchi A, Babu RC, Price A, McNally KL, Henry A (2015) Environmental response and genomic regions correlated with rice root growth and yield under drought in the OryzaSNP panel across multiple study systems. PLoS One 10:e0124127

Wang H, Yamauchi A (2006) Growth and function of roots under abiotic stress soils. In: Huang B (ed) Plant environment interactions, 3rd edn. CRC Press/, Taylor and Francis Group, LLC, New York, NY, pp 271–320

Wang H, Inukai Y, Kamoshita A, Wade LJ, Siopongco JDLC, Nguyen H, Yamauchi A (2005) QTL analysis on plasticity in lateral root development in response to water stress in the rice plant. In: Toriyama K, Heong KL, Hardy B (eds) Rice is Life: Scientific Perspectives for the 21st Century. The Proceedings of the World Rice Research Conference, Tsukuba, Japan. International Rice Research Institute/Japan International Research Center for Agricultural Sciences, Los Baños, Philippines/Tsukuba, Japan, pp 464–469

Wang H, Inukai Y, Yamauchi A (2006) Root Development and Nutrient Uptake. Crit Rev Plant Sci 25:279–301

Watanabe Y, Kabuki T, Kakehashi T, Kano-Nakata M, Mitsuya S, Yamauchi A (2020) Morphological and histological differences among three types of component roots and their differential contribution to water uptake in the rice root system. Plant Prod Sci 23:191–201. https://doi.org/10.1080/1343943X.2020.1730701

Wiengweera A, Greenway H, Thomson CJ (1997) The use of agar nutrient solution to simulate lack of convection in waterlogged soils. Ann Bot 80:115–123

Wissusa M, Kretzschmar T, Rose TJ (2016) From promise to application: root traits for enhanced nutrient capture in rice breeding. J Exp Bot 67(12):3605–3615

Xu W, Cui K, Xu A, Nie L, Haung J, Peng S (2015) Drought stress condition increases root to shoot ratio via alteration of carbohydrate partitioning and enzymatic activity in rice seedlings. Acta Physiol Plant 37:9

Yamauchi A (2003) Ideotype of root system. In Japanese. In: Morita S (ed) Principle of root design. Youkendo, Tokyo, pp 10–17

Yamauchi A, Pardales JR Jr, Kono Y (1996) Root system structure and its relation to stress tolerance. In: Ito O, Katayama K, Johansen C, Kumar Rao JVDK, Adu-Gyamfi JJ, Rego TJ (eds) Roots and nitrogen in cropping systems of the semi-arid tropics. JIRCAS Publication, Tsukuba, pp 211–234

Yamauchi T, Colmer TD, Pedersen O, Nakazono M (2018) Regulation of root traits for internal aeration and tolerance to soil waterlogging-flooding stress. Plant Physiol 76(2):1118–1130. https://doi.org/10.1104/pp.17.01157

Yamauchi T, Pedersen O, Nakazono M, Tsutsumi N (2021a) Key root traits of Poaceae for adaptation to soil water gradients. New Phytol 229(6):3133

Yamauchi Y, Noshita K, Tsutsumi N (2021b) Climate-smart crops: key root anatomical traits that confer flooding tolerance. Breed Sci 71:51–61. https://doi.org/10.1270/jsbbs.20119

Zhang J, Zheng HG, Aarti A, Pantuwan G, Nguyen TT, Tripathy JN, Sarial AK, Robin S, Babu RC, Nguyen BD, Sarkarung S, Blum A, Nguyen HT (2001) Locating genomic regions associated with components of drought resistance in rice: comparative mapping within and across species. Theor Appl Genet 103:19–29

Zhu J, Zhang C, Lynch JP (2010) The utility of phenotypic plasticity of root hair length for phosphorus acquisition. Funct Plant Biol 37:313–322

Chapter 4
Regulation of Root Tissue Size and Adaptations to Hypoxia

Takaki Yamauchi

4.1 Introduction

Soil flooding is defined as a root zone being filled with excess water that restricts oxygen diffusion from the air into the soil (Sasidharan et al. 2017). As the slow diffusion rate in flooded soils results in an inadequate supply of oxygen into submerged tissues, root cells of non-wetland plants are damaged by severe hypoxia or anoxia (Colmer and Voesenek 2009). Plants have no active oxygen dispersal mechanisms; thus, oxygen movement within plants under flooding mostly depends on oxygen diffusion (Armstrong and Armstrong 2014). Lysigenous aerenchyma is a large internal gas space created by the death and lysis of the cells and provides a low-resistance gas-phase diffusion of oxygen along roots (Jackson and Armsrtong 1999). Many plants including the families of Poaceae and Fabaceae can form lysigenous aerenchyma in the root cortex (Justin and Armsrtong 1987). However, the time course and extent of its formation largely differ among the species (Colmer and Voesenek 2009; Yamauchi et al. 2018).

Roots mainly consist of three concentric cell layers; from the outside to inside are the epidermis, cortex, and stele (Coudert et al. 2010; Dolan et al. 1993). The cortex includes the endodermis and hypodermis (exodermis) and in some cases the sclerenchyma (Fig. 4.1) (Lux et al. 2004). The stele contains xylem vessels and phloem tissue, which have a role in water and nutrient transport (Petricka et al. 2012). The main root of *Arabidopsis* (*Arabidopsis thaliana*) has a single cortex cell layer produced by a single asymmetric cell division of cortex/endodermis initial cells, if the endodermis (a part of the cortex) and middle cortex (only detected in late stage) are not considered as cortical cells (Dolan et al. 1993; Lux et al. 2004; Paquette and Benfey 2005). On the other hand, the sequential cell divisions of

T. Yamauchi (✉)
Bioscience and Biotechnology Center, Nagoya University, Nagoya, Japan
e-mail: atkyama@agr.nagoya-u.ac.jp

J.-I. Sakagami, M. Nakazono (eds.), *Responses of Plants to Soil Flooding*,
https://doi.org/10.1007/978-981-99-9112-9_4

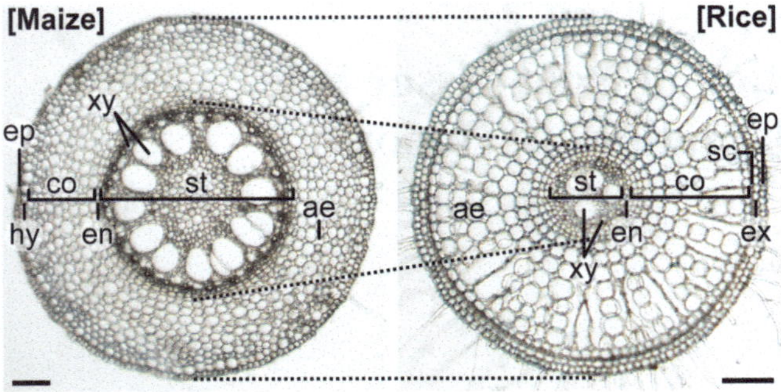

Fig. 4.1 Root tissue structures in maize and rice. Root cross-sections at 40 mm from the adventitious root tips of maize (inbred line B73, left) and rice (cv. Nipponbare, right) grown under aerated nutrient solutions. Bars = 100 μm. Dotted lines show that maize and rice have a larger proportion of stele and cortex, respectively. Wetland species, such as rice, generally have clear sclerenchyma cell layer(s) with a thick cell wall. In a strict sense, the endodermis, sclerenchyma, and hypodermis/exodermis are parts of cortex. *ep* epidermis, *hy* hypodermis, *co* cortex, *en* endodermis, *st* stele, *xy* xylem, *ae* aerenchyma, *ex* exodermis, *sc* sclerenchyma

cortex/endodermis initial cells generate several cortex cell layers in rice roots (Kamiya et al. 2003a). The roots of most plant species, e.g., rice and maize, have more than three cortex cell layers (Fig. 4.1) (Justin and Armsrtong 1987; Yamauchi et al. 2019a).

Root anatomical traits may determine the balance between the water and oxygen transport within roots (Yamauchi et al. 2021a). Root thickness has been considered to be an essential trait for tolerance to soil flooding (Colmer 2003). Thick roots enhance longitudinal oxygen diffusion by reducing the resistance to gas diffusion (Colmer 2003). Indeed, wetland species generally have thicker and more porous roots than non-wetland species (Visser et al. 2000). Underlying mechanism of the benefit of thick root systems can be explained by a surface area-to-volume ratio (SA/V); the thick roots with a smaller SA/V is less prone to lose oxygen from the root surface (Pedersen et al. 2021). Moreover, the proportion of the volumes of cortex and aerenchyma should correlate to the SA/V, given that the porous cortex is the main path for oxygen diffusion (Colmer 2003; Yamauchi et al. 2021a). However, these ideas also do not fully explain why wetland plants can grow well under soil flooding, as a small proportion of stele and xylem may reduce the water and nutrient uptakes from the soils (Yamauchi et al. 2021a). To overcome this trade-off, rice produces more adventitious roots than wheat and maize (Yamauchi et al. 2019a).

In this chapter, I provide an overview of the mechanisms of root radial patterning, which eventually determine the root-tissue sizes. Moreover, I summarize some recent advances in understanding the adaptive significance of root-tissue size to soil flooding and discuss the possible mechanisms underlying root-tissue size control in plants based on the fundamental regulation of root radial patterning.

4.2 Radial Patterning of Roots

The stem cells at the root apex, i.e., root apical meristem (RAM), provide almost all of the cells that compose the roots (Fig. 4.2) (Stromann and Stahl 2021). The RAM is located just behind the root cap, which consists of the columella and the lateral cap (Stromann and Stahl 2021). In the primary root of *Arabidopsis* (*Arabidopsis thaliana*), the RAM has four rarely dividing quiescent center (QC) cells (Stromann and Stahl 2021). The initial cells of each root tissue are then produced by the asymmetric divisions of the QC cells (Dolan et al. 1993; Petricka et al. 2012). The radial positions of the initial cells are fixed throughout the root growth, and thus concentrically organized cell lineages, i.e., tissue layers, are established (Dolan et al. 1993; Petricka et al. 2012). The primary root of *Arabidopsis* has a single cortical cell layer (Dolan et al. 1993; Lux et al. 2004). On the other hand, seminal roots of most plant species, such as rice, maize, and soybean (*Glycine max* L.), have multiple cortical cell layers (Kamiya et al. 2003b; Dong et al. 2021; Ortiz-Ramírez et al. 2021). More than four mitotically inactive cells are identified in the rice RAM (Ni et al. 2014), even though the number of QC cells is still controversial (Tanaka et al. 2023). As thick maize root has hundreds more QC cells (Jiang et al. 2003), a large number of QC cells can be required to maintain the complex and large root organization (Jiang and Feldman 2005; Stromann and Stahl 2021).

Unlike eudicots with tap root systems, monocots have fibrous root systems consisting of a few seminal roots and multiple shoot-born adventitious roots. In rice, the development of adventitious (crown) roots starts with an initial cell generated by the periclinal division of the innermost ground meristem in the stem nodes (Itoh et al. 2005; Viana et al. 2022; Tanaka et al. 2023). The initial cells are then divided anticlinally and periclinally to establish the adventitious root primordia, which consist

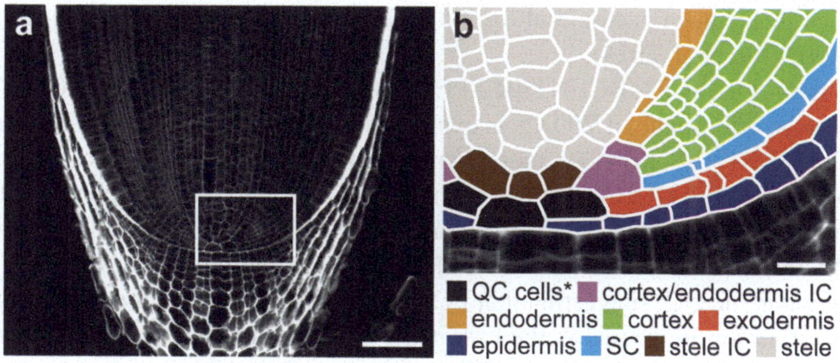

Fig. 4.2 The root apical meristem (RAM) of rice adventitious roots. (**a**) Confocal micrograph of the RAM. Scale bar = 50 μm. (**b**) The enlarged view of the RAM. Scale bar = 10 μm. The cells consisting RAM region are distinguished by different colors. The putative quiescent center (QC) cells are indicated in black color, as the exact number and extent of QC cells in rice are still controversial. The putative cortex/endodermis initial cells (ICs) and stele ICs are also defined according to the previous reports (Coudert et al. 2010; Ni et al. 2014; Strotmann and Stahl 2021). *SC* sclerenchyma

of epidermis, endodermis, stele, and root cap initials (Itoh et al. 2005; Viana et al. 2022; Tanaka et al. 2023). The cortex/endodermis initial cells are divided several times in an asymmetrical and periclinal manner to generate multiple cortical cell layers in rice (Kamiya et al. 2003a). Subsequently, xylem vessels are generated in the stele, and the root primordia vascularly connect to the stem node in rice (Itoh et al. 2005; Viana et al. 2022; Tanaka et al. 2023). Once the root primordia are established, the radial pattern of the root cells is maintained continuously throughout the root growth (Itoh et al. 2005; Viana et al. 2022; Tanaka et al. 2023).

4.3 Molecular Mechanisms of the Radial Patterning of Roots

In *Arabidopsis*, QC cells produce a signal for maintaining stem cell identity in the RAM (Petricka et al. 2012). WUSCHEL-RELATED HOMEOBOX5 (WOX5) is expressed in the QC and prevents adjacent stem cell differentiation (Sarkar et al. 2007). Restriction of the AtWOX5 expression to the QC cells is regulated through the function of the CLAVATA3/EMBRYO SURROUNDING REGION40 peptide, which interaction determines the extent of stem cells (Stahl et al. 2009). Two GRAS family transcription factors SHORT-ROOT (SHR) and SCARECROW (SCR) control an asymmetric division of the cortex/endodermis initial cells (Di Laurenzio et al. 1996; Helariutta et al. 2000; Nakajima et al. 2001; Petricka et al. 2012). AtSHR is expressed in the stele and triggers asymmetric cell divisions of the stelar cells (Cui et al. 2007). Moreover, the AtSHR protein moves into the adjacent cell layer and stimulates the expression of AtSCR in the cortex/endodermis initial cell (Cui et al. 2007). AtSCR has a role in the asymmetric cell division of the cortex/endodermis initial cell and also functions in the restriction of the radial AtSHR protein movement to specify the endodermis (Cui et al. 2007; Wu et al. 2014). This local feedback mechanism is essential for the root radial patterning in *Arabidopsis* (Cui et al. 2007; Wu et al. 2014).

In rice, QUISCENT-CENTER-SPECIFIC HOMEOBOX (QHB)/OsWOX5, which is the orthologue of WOX5, is expressed in the QC region (Fig. 4.3) (Kamiya et al. 2003b; Ni et al. 2014). As more than four mitotically inactive cells are detected concomitantly with QHB/OsWOX5 promoter activity in the RAMs (Ni et al. 2014), rice may have multiple QC cells (Stromann and Stahl 2021). The application of the FON2-LIKE CLE PROTEIN2 (FCP2) peptide suppresses the QHB/OsWOX5 expression in the QC, suggesting that FCP2-QHB/OsWOX5 signaling controls root cell proliferation in rice as is the case with CLE-WOX signaling in *Arabidopsis* (Chu et al. 2013). One of two rice SCR orthologues *OsSCR2* is expressed in the QC and cortex/endodermis initial cells, and the *OsSCR2* expression in the daughter cortical cell is decreased after the periclinal division of the cortex/endodermis initial cells (Fig. 4.3) (Kamiya et al. 2003a; Ni et al. 2014). The overexpressors of *AtSHR* and *OsSHR2* produce numerous numbers of cortical cell layers in rice roots, suggesting that SHRs function in determining the number of cortical cell layers (Henry et al. 2017).

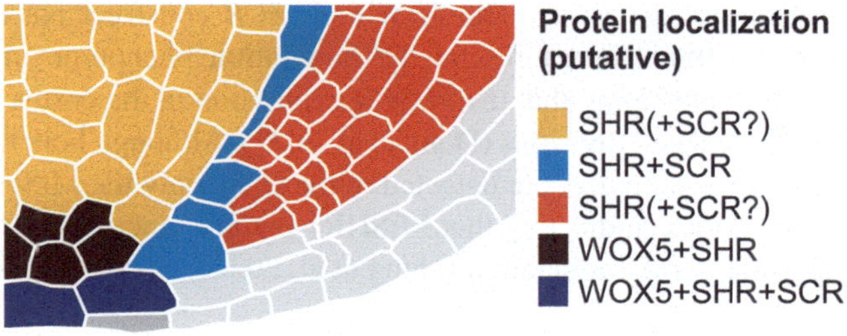

Fig. 4.3 The putative localizations of the proteins that determine the root radial patterns in Poaceae. In maize, the *SHR* transcripts are detected in the stele and endodermis, and the SHR proteins move into the cortex (Ortiz-Ramírez et al. 2021). SHRs stimulate the asymmetric divisions of cortex/endodermis initial cells and possibly also the cortical cells in rice and maize (Henry et al. 2017; Ortiz-Ramírez et al. 2021). The *SCR* transcripts are detected in the endodermis of rice and maize, and SCRs function in the specification of the endodermis (Kamiya et al. 2003a; Ni et al. 2013; Ortiz-Ramírez et al. 2021). In maize, the SCR protein moves into the stelar cells (Ortiz-Ramírez et al. 2021). In rice, the *QHB/WOX5* transcript is detected in several cells located in the center of the root apical meristem (Kamiya et al. 2003b; Ni et al. 2013; Chu e al. 2014). QHB/WOX5 determines the region of the quiescent center (QC) and initial cells. Both SCR and SHR transcripts are detected in the QC of rice and maize (Kamiya et al. 2003a; Ni et al. 2013; Ortiz-Ramírez et al. 2021)

A recent finding in maize, which has multiple cortical cell layers, shows differences in the SHR-SCR-dependent radial patterning with that in *Arabidopsis* (Ortiz-Ramírez et al. 2021). In maize, the mRNAs of three *ZmSHR* genes are expressed in the QC, stele, and endodermis, but ZmSHR1 protein moves at least eight cell layers into the cortex (Fig. 4.3) (Ortiz-Ramírez et al. 2021). On the other hand, the ZmSCR1 protein moves into the stele in the opposite direction from SHR, even though the mRNA of the *ZmSCR* is detected only in the endodermis (Fig. 4.3) (Ortiz-Ramírez et al. 2021). As the knockout mutant of *ZmSHRs* shows a significant reduction in the number of cortical cell layers, SHR proteins in the cortex are essential to establish multiple cortical cell layers in maize (Ortiz-Ramírez et al. 2021). The knockout mutant of two *SHR* genes in *Setaria virdis* also shows a significant reduction in the number of cortical cell layers, suggesting that the SHR function is conserved at least in the subfamily Panicoideae (Ortiz-Ramírez et al. 2021). Interestingly, the heterologous expression of the rice *OsSHR2* in *Arabidopsis* results in the hypermobility of the OsSHR2 protein into the cortex (Fig. 4.3) (Cui et al. 2007).

Another line of evidence showing the SHR-SCR-dependent control of the multiple cortical cell layers comes from a study for legume symbiosis (Dong et al. 2021). During the nodule organogenesis in hairy roots of barrel medic (*Medicago truncatula*), the cortical cells are reprogrammed and initiate cell divisions (Dong et al. 2021). Interestingly, this de novo cortical cell division is sufficiently activated by the ectopic SHR and SCR expression. Unlike in *Arabidopsis*, rice, and maize (Kamiya et al. 2003a, Wu et al. 2014; Ortiz-Ramírez et al. 2021), the mRNA of the *MtSCR* gene is expressed not only in the endodermis but also in the cortical cells of barrel medic (Dong et al. 2021). The *MtSCR*::*MtSCR* construct rescues but the *AtSCR*::*MtSCR* construct cannot rescue the nodulation phenotype in the *Mtscr* mutant background, indicating that the *MtSCR* expression in the cortex is essential for the de novo cortical cell divisions (Dong et al. 2021). Although both the *MtSHR1* and *MtSHR2* promoter activities are detected in the stele, only the MtSHR1 protein can move into the cortex, indicating that MtSHR1 mainly functions in the de novo cortical cell division during nodulation (Dong et al. 2021).

4.4 Root Tissue-Size Controls for Adaptation to Soil Flooding

Plants with a large cortex and small stele proportion in the roots better adapt to soil flooding (Colmer and Greenway 2011). In maize roots, the oxygen level is higher in the porous cortex and much lower in the stele with a higher density of cells (Armstrong et al. 1994; Gibbs et al. 1998). Oxygen deficiency in the stelar cells of maize and barley (*Hordeum vulgare*) strongly restricts the loading of essential ions into the xylem vessels (Gibbs et al. 1998; Kotula et al. 2015). Thus, having a small stele is advantageous for plants to reduce the risk of anoxia of the stelar cells by the oxygen supply through a large porous cortex and to maintain the efficiency of nutrient uptakes from the roots under soil flooding (Armstrong et al. 2019). Indeed, the roots of several wetland species have a smaller stele proportion when compared with those of dryland grass species (Justin and Armstrong 1987; McDonald et al. 2002). Moreover, a smaller stele proportion always associates with a higher root porosity (McDonald et al. 2002), which is calculated as the gas volume per root volume (Justin and Armstrong 1987).

Recently, this idea has been supported by the use of the root anatomical indices representing the ratio of root-tissue areas; cortex-to-stele ratio (CSR) and aerenchyma-to-cortex ratio (ACR) enable to identify key root anatomical features for the adaptation of plants to various soil water content (Yamauchi et al. 2021b). Although the area of each root tissue of the Poaceae species correlates to the plant size, CSR and ACR correlate with the soil water content of their natural habitats (Yamauchi et al. 2021b). High ACR could reduce metabolic costs both under drought and soil flooding (Colmer 2003; Zhu et al. 2010; Lynch 2018) and also enhance oxygen diffusion in the cortex under soil flooding (Colmer 2003). High CSR could increase the space to form aerenchyma and intercellular spaces under soil flooding (Yamauchi et al. 2019a). Moreover, high CSR can lower the risk of stele anoxia, which prevents nutrient transport from the soils (Armstrong et al. 2019). A modeling approach predicts that increasing CSR is beneficial for the maximum lengths achieved by the roots under soil flooding, and the benefit could be more substantial for the species having a small CSR (e.g., maize and wheat) than those having a large enough CSR (e.g., rice) (Pedersen et al. 2021).

4.5 Root Tissue-Size Plasticity in Response to Soil Flooding

The plasticity of lysigenous aerenchyma formation under soil flooding has been widely studied for a long time (Jackson and Armstrong 1999; Colmer 2003; Colmer and Voesenek 2009; Voesenek and Bailey-Serres 2015). Roots of wetland species developmentary form aerenchyma under aerobic conditions without any stressors, such as nutrient deficiency, but most of the dryland (upland) species hardly form aerenchyma under aerobic conditions (Colmer and Voesenek 2009). Under low-oxygen conditions, both wetland and dryland species increase aerenchyma formation (Colmer and Voesenek 2009). On the other hand, a recent study using the modeling-based age-dependent analysis showed that adventitious roots of rice more rapidly induce aerenchyma formation under low-oxygen conditions than those of maize (Yamauchi and Nakazono 2022b). This rapid induction could contribute to aerenchyma formation at the more apical part of the adventitious roots of rice to enhance oxygen diffusion to the root tips (Yamauchi and Nakazono 2022b). Although many plant species form lysigenous aerenchyma (Justin and Armstrong 1987), its plasticity remains unclear except for plants of the family Poaceae.

The ratio of the cortex area and stele area is essential for the tolerance to flooding and drought, respectively (Yamauchi et al. 2021a). This simultaneously implies that a trade-off underlies root adaptation to flooding and drought, as an increase in the cortex proportion, resulting in a decrease in the stele proportion and vice versa. Nevertheless, plants should respond to soil water fluctuation due to the random precipitation in nature. Indeed, rice, maize, and wheat roots all show the plasticity of CSR and ACR in response to low-oxygen conditions (Yamauchi et al. 2019a). Although a wetland species of rice has higher CSR and ACR both under aerobic and low-oxygen conditions than dryland (upland) species of maize and wheat, the CSR and ACR in all three species are significantly increased under low-oxygen conditions (Fig. 4.4) (Yamauchi et al. 2019a, 2021a). Interestingly, continuous rainfall increases CSR and ACR in the roots of four wild Poaceae species (Yamauchi et al. 2021b). Together, root tissue-size plasticity is the key for plants to overcome the trade-off during adaptation to intermittent flooding.

4.6 Perspective: Mechanisms Underlying Root Tissue-Size Plasticity

The physiological and molecular mechanisms underlying the plasticity of lysigenous aerenchyma formation have been intensively studied in rice and maize (Drew et al. 2000; Voesenek and Bailey-Serres 2015; Yamauchi and Nakazono 2022a). In rice roots, aerenchyma is developmentary formed through the AUXIN RESPONSE FACTOR- and AUX/IAA-dependent auxin signaling (Yamauchi et al. 2019b). On the other hand, ethylene and reactive oxygen species (ROS) signaling is required for

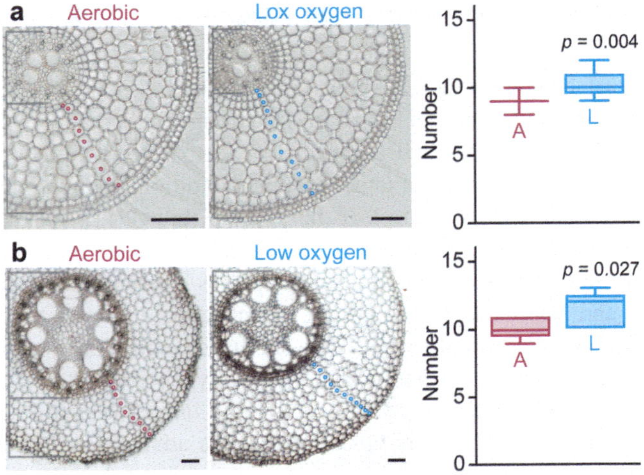

Fig. 4.4 Number of the cortical cell layers in rice and maize adventitious roots. (**a**) Root cross-sections at 20 mm from the root tips of rice (cv. Nipponbare) under aerobic and low-oxygen conditions. Numbers of cortical cell layers (without endodermis, epidermis, exodermis, and sclerenchyma) under aerobic (A) and low-oxygen (L) conditions. (**b**) Root cross-sections at 20 mm from the root tips of maize (inbred line B73) under aerobic and low-oxygen conditions. Numbers of cortical cell layers (without endodermis and epidermis) under aerobic (A) and low-oxygen (L) conditions. Boxplots show the median (horizontal lines), 25th to 75th percentiles (edges of the boxes), and minimum to maximum (edges of the whiskers) ($n = 9$). P-values (two-sample t-test) between aerobic and low-oxygen conditions are indicated above the boxplots. Bars = 100 μm. Dots in the photographs show the alignments of the cortical cells

the induction of aerenchyma formation in rice, maize, and wheat (Yamauchi et al. 2018; Yamauchi and Nakazono 2022a). During the induction of aerenchyma formation in rice roots, RESPIRATORY BURST OXIDASE HOMOLOG (RBOH)-mediated ROS production is stimulated by the CALCIUM-DEPENDENT PROTEIN KINASE (CDPK)-dependent phosphorylation (Yamauchi et al. 2017). The treatments of an RBOH inhibitor or a calcium chelator prevent aerenchyma formation in maize and wheat (Drew et al. 2000; Yamauchi et al. 2018), suggesting that RBOH and CDPK commonly regulate inducible aerenchyma formation in many plant species.

So far, the physiological and molecular mechanisms underlying cortex expansion in response to low-oxygen conditions remain unknown. However, some suggestive studies have been reported on cortex-size plasticity. SCR-SHR-dependent asymmetric cell divisions are required to control the numbers of cortical cell layers at least in maize and barrel medic (Dong et al. 2021; Ortiz-Ramírez et al. 2021). These studies clearly showed that the extent and intensity of SCR and/or SHR proteins determine the duration of cortex/endodermis cell and cortical cell divisions (Dong et al. 2021; Ortiz-Ramírez et al. 2021). In rice and maize roots, the number of cortical cell layers increased under low-oxygen conditions (Fig. 4.4) (Yamauchi et al. 2019a). As the overexpression of *OsSHR2* greatly stimulates the asymmetric divisions of the cortical cells (Henry et al. 2017), the induction of the *SHR* expression, which might be accompanied by the *SCR* expression, underlies the root-cortex expansion in response to low-oxygen conditions under soil flooding. Another possibility is that the extent of SCR and/or SHR proteins in the roots is expanded under soil flooding.

The maintenance of the root radial patterning is regulated mainly through balancing two phytohormones, i.e., auxin and cytokinin (Viana et al. 2022; Tanaka et al. 2023). The CROWN ROOTLESS1 (CRL1) is regulated through the ARF-dependent auxin signaling (Inukai et al. 2005). As CRL1 activates the *QHB/OsWOX5* expression (Lavarenne et al. 2019), CRL1 is a key regulator for adventitious root initiation and maintenance (Inukai et al. 2005; Viana et al. 2022; Tanaka et al. 2023). Although QHB/OsWOX5 has a role not only in QC maintenance but also in xylem development (Chu et al. 2013), its interaction with *SCR* and/or *SHR* is not fully understood. As auxin is required for the induction of aerenchyma formation in rice roots under low-oxygen conditions (Yamauchi et al. 2020), auxin accumulation and/or signaling might be affected by ethylene and/or oxygen deficiency. Taken together, auxin-dependent controls of WOX5, SCR, and/or SHR expression may underly root-tissue size controls in plants under soil flooding. Further studies are required to better understand root-tissue size plasticity in response to soil flooding.

Acknowledgments I thank Dr. Katsutoshi Tsuda for kindly providing the photograph of root apical meristem in rice. This work was partly supported by the Grant-in-Aid for Transformative Research Areas (A) (MEXT KAKENHI Grant Numbers, 23H04195) and JSPS KAKENHI 22H02308.

References

Armstrong W, Armstrong J (2014) Plant internal oxygen transport (diffusion and convection) and measuring and modelling oxygen gradients. Plant Cell Monogr 21:267–297

Armstrong W, Strange ME, Cringle S, Beckett PM (1994) Microelectrode and modelling study of oxygen distribution in roots. Ann Bot 74:287–299

Armstrong W, Beckett PM, Colmera TD, Setter TL, Greenway H (2019) Tolerance of roots to low oxygen: 'anoxic' cores, the phytoglobin-nitric oxide cycle, and energy or oxygen sensing. J Plant Physiol 239:92–108

Chu H, Liang W, Li J, Hong F, Wu Y, Wang L, Wang J, Wu P, Liu C, Zhang Q, Xu J, Zhang D (2013) A CLE-WOX signalling module regulates root meristem maintenance and vascular tissue development in rice. J Exp Bot 64:5359–5369

Colmer TD (2003) Long-distance transport of gases in plants: a perspective on internal aeration and radial oxygen loss from roots. Plant Cell Environ 26:17–36

Colmer TD, Greenway H (2011) Ion transport in seminal and adventitious roots of cereals during O deficiency. J Exp Bot 62:39–57

Colmer TD, Voesenek LACJ (2009) Flooding tolerance: suites of plant traits in variable environments. Funct Plant Biol 36:665–681

Coudert Y, Périn C, Courtois B, Khong NG, Gantet P (2010) Genetic control of root development in rice, the model cereal. Trends Plant Sci 15:219–226

Cui H, Levesque MP, Vernoux T, Jung JW, Paquette AJ, Gallagher KL, Wang JY, Blilou I, Scheres B, Benfey PN (2007) An evolutionarily conserved mechanism delimiting SHR movement defines a single layer of endodermis in plants. Science 316:421–425

Di Laurenzio L, Wysocka-Diller J, Malamy JE, Pysh L, Helariutta Y, Freshour G, Hahn MG, Feldmann KA, Benfey PN (1996) The *SCARECROW* gene regulates an asymmetric cell division that is essential for generating the radial organization of the Arabidopsis root. Cell 86:423–433

Dolan L, Janmaat K, Willemsen V, Linstead P, Poethig S, Roberts K, Scheres B (1993) Cellular organisation of the Arabidopsis thaliana root. Development 119:71–84

Dong W, Zhu Y, Chang H, Wang C, Yang J, Shi J, Gao J, Yang W, Lan L, Wang Y, Zhang X, Dai H, Miao Y, Xu L, He Z, Song C, Wu S, Wang D, Yu N, Wang E (2021) An SHR-SCR module specifies legume cortical cell fate to enable nodulation. Nature 589:586–590

Drew MC, He C-J, Morgan PW (2000) Programmed cell death and aerenchyma formation in roots. Trends Plant Sci 5:123–127

Gibbs J, Turner DW, Armstrong W, Darwent MJ, Greenway H (1998) Response to oxygen deficiency in primary maize roots. I. Development of oxygen deficiency in the stele reduces radial solute transport to the xylem. Aust J Plant Physiol 25:745–758

Helariutta Y, Fukaki H, Wysocka-Diller J, Nakajima K, Jung J, Sena G, Hauser M-T, Benfey PN (2000) The *SHORT-ROOT* gene controls radial patterning of the Arabidopsis root through radial signaling. Cell 101:555–567

Henry S, Dievart A, Divol F, Pauluzzi G, Meynard D, Swarup R, Wu S, Gallagher KL, Périn C (2017) *SHR* overexpression induces the formation of supernumerary cell layers with cortex cell identity in rice. Dev Biol 425:1–7

Inukai Y, Sakamoto T, Ueguchi-Tanaka M, Shibata Y, Gomi K, Umemura I, Hasegawa Y, Ashikari M, Kitano H, Matsuoka M (2005) *Crown rootless1*, which is essential for crown root formation in rice, is a target of an AUXIN RESPONSE FACTOR in auxin signaling. Plant Cell 17:1387–1396

Itoh J, Nonomura K, Ikeda K, Yamaki S, Inukai Y, Yamagishi H, Kitano H, Nagato Y (2005) Rice plant development: from zygote to spikelet. Plant Cell Physiol 46:23–47

Jackson MB, Armsrtong W (1999) Formation of aerenchyma and the proceses of plant ventilation in relation to soil flooding and submergence. Plant Biol 1:274–287

Jiang K, Feldman LJ (2005) Regulation of root apical meristem development. Annu Rev Cell Dev Biol 21:485–509

Jiang K, Meng YL, Feldman LJ (2003) Quiescent center formation in maize roots is associated with an auxin-regulated oxidizing environment. Development 130:1429–1438

Justin SHFW, Armstrong W (1987) The anatomical characteristics of roots and plant response to soil flooding. New Phytol 106:465–495

Kamiya N, Itoh J, Morikami A, Nagato Y, Matsuoka M (2003a) The *SCARECROW* gene's role in asymmetric cell divisions in rice plants. Plant J 36:45–54

Kamiya N, Nagasaki H, Morikami A, Sato Y, Matsuoka M (2003b) Isolation and characterization of a rice *WUSCHEL*-type homeobox gene that is specifically expressed in the central cells of a quiescent center in the root apical meristem. Plant J 35:429–441

Kotula L, Clode PL, Striker GG, Pedersen O, Läuchli A, Shabala S, Colmer TD (2015) Oxygen deficiency and salinity affect cellspecific ion concentrations in adventitious roots of barley (Hordeum vulgare). New Phytol 208:1114–1125

Lavarenne J, Gonin M, Guyomarc'h S, Rouster J, Champion A, Sallaud C, Laplaze L, Gantet P, Lucas M (2019) Inference of the gene regulatory network acting downstream of CROWN ROOTLESS 1 in rice reveals a regulatory cascade linking genes involved in auxin signaling, crown root initiation, and root meristem specification and maintenance. Plant J 100:954–968

Lux A, Luxová M, Abe J, Morita S (2004) Root cortex: structural and functional variability and responses to environmental stress. Root Res 13:117–131

Lynch JP (2018) Rightsizing root phenotypes for drought resistance. J Exp Bot 69:3279–3292

McDonald MP, Galwey NW, Colmer TD (2002) Similarity and diversity in adventitious root anatomy as related to root aeration among a range of wetland and dryland grass species. Plant Cell Environ 25:441–451

Nakajima K, Sena G, Nawy T, Benfey PN (2001) Intercellular movement of the putative transcription factor SHR in root patterning. Nature 413:307–311

Ni J, Shen Y, Zhang Y, Wu P (2014) Definition and stabilisation of the quiescent centre in rice roots. Plant Biol 16:1014–1019

Ortiz-Ramírez C, Guillotin B, Xu X, Rahni R, Zhang S, Yan Z, Araujo PCD, Demesa-Arevalo E, Lee L, Van Eck J, Gingeras TR, Jackson D, Gallagher KL, Birnbaum KD (2021) Science 374:1247–1252

Paquette AJ, Benfey PN (2005) Maturation of the ground tissue of the root is regulated by gibberellin and *SCARECROW* and requires *SHORT-ROOT*. Plant Physiol 138:636–640

Pedersen O, Sauter M, Colmer TD, Nakazono M (2021) Regulation of root adaptive anatomical and morphological traits during low soil oxygen. New Phytol 229:42–49

Petricka JJ, Winter CM, Benfey PN (2012) Control of Arabidopsis root development. Annu Rev Plant Biol 63:563–590

Sarkar AK, Luijten M, Miyashima S, Lenhard M, Hashimoto T, Nakajima K, Scheres B, Heidstra R, Laux T (2007) Conserved factors regulate signalling in Arabidopsis thaliana shoot and root stem cell organizers. Nature 446:811–814

Sasidharan R, Bailey-Serres J, Ashikari M, Atwell BJ, Colmer TD, Fagerstedt K, Fukao T, Geigenberger P, Hebelstrup KH, Hill RD, Holdsworth MJ, Ismail AM, Licausi F, Mustroph A, Nakazono M, Pedersen O, Perata P, Sauter M, Shih MC, Sorrell BK, Striker GG, van Dongen JT, Whelan J, Xiao S, Visser EJW, Voesenek LACJ (2017) Community recommendations on terminology and procedures used in flooding and low oxygen stress research. New Phytol 214:1403–1407

Stahl Y, Wink RH, Ingram GC, Simon R (2009) A signaling module controlling the stem cell niche in Arabidopsis root meristems. Curr Biol 19:909–914

Stromann VI, Stahl Y (2021) At the root of quiescence: function and regulation of the quiescent center. J Exp Bot 72:6716–6726

Tanaka W, Yamauchi T, Tsuda K (2023) Genetic basis controlling rice plant architecture and its modification for breeding. Breed Sci 73:3–45

Viana WG, Scharwies JD, Dinneny JR (2022) Deconstructing the root system of grasses through an exploration of development, anatomy and function. Plant Cell Environ 45:602–619

Visser EJW, Colmer TD, Blom CWPM, Voesenek LACJ (2000) Changes in growth, porosity, and radial oxygen loss from adventitious roots of selected mono- and dicotyledonous wetland species with contrasting types of aerenchyma. Plant Cell Environ 23:1237–1245

Voesenek LACJ, Bailey-Serres J (2015) Flood adaptive traits and processes: an overview. New Phytol 206:57–73

Wu S, Lee CM, Hayashi T, Price S, Divol F, Henry S, Pauluzzi G, Perin C, Gallagher KL (2014) A plausible mechanism, based upon SHORT-ROOT movement, for regulating the number of cortex cell layers in roots. Proc Natl Acad Sci USA 111:16184–16189

Yamauchi T, Nakazono M (2022a) Mechanisms of lysigenous aerenchyma formation under abiotic stress. Trends Plant Sci 27:13–15

Yamauchi T, Nakazono M (2022b) Modeling-based age-dependent analysis reveals the net patterns of ethylene-dependent and -independent aerenchyma formation in rice and maize roots. Plant Sci 321:111340

Yamauchi T, Yoshioka M, Fukazawa A, Mori H, Nishizawa NK, Tsutsumi N, Yoshioka Nakazono M (2017) An NADPH oxidase RBOH functions in rice roots during lysigenous aerenchyma formation under oxygen-deficient conditions. Plant Cell 29:775–790

Yamauchi T, Colmer TD, Pedersen O, Nakazono M (2018) Regulation of root traits for internal aeration and tolerance to soil waterlogging-flooding stress. Plant Physiol 176:1118–1130

Yamauchi T, Abe F, Tsutsumi N, Nakazono M (2019a) Root cortex provides a venue for gas-space formation and is essential for plant adaptation to waterlogging. Front Plant Sci 10:259

Yamauchi T, Tanaka A, Inahashi H, Nishizawa NK, Tsutsumi N, Inukai Y, Nakazono M (2019b) Fine control of aerenchyma and lateral root development through AUX/IAA- and ARF-dependent auxin signaling. Proc Natl Acad Sci U S A 116:20770–20775

Yamauchi T, Tanaka A, Tsutsumi N, Inukai Y, Nakazono M (2020) A role for auxin in ethylene-dependent inducible aerenchyma formation in rice roots. Plan Theory 9:610
Yamauchi T, Noshita K, Tsutsumi N (2021a) Climate-smart crops: key root anatomical traits that confer flooding tolerance. Breed Sci 71:51–61
Yamauchi T, Pedersen O, Nakazono M, Tsutsumi N (2021b) Key root traits of Poaceae for adaptation to soil water gradients. New Phytol 229:3133–3140
Zhu J, Brown KM, Lynch JP (2010) Root cortical aerenchyma improves the drought tolerance of maize (Zea mays L.). Plant Cell Environ 33:740–749

Part III
Reactions in Above Ground Stems and Leaves

Chapter 5
Flood Avoidance Mechanism Via Shoot Elongation and Photosynthesis in Rice Plants

Jun-Ichi Sakagami

5.1 Introduction

Rice is one of the few crops that can tolerate waterlogging; however, its growth declines with prolonged complete submergence. In the future, it is predicted that the number of areas subjected to flooding will increase due to global warming and the frequency of torrential or specific rains. Consequently, it is extremely important that we develop flood damage countermeasures for rice cultivation in flood-prone areas. One of the most important strategies for this is improving flood resistance characteristics in rice plants. In general, rice plants exhibit submergence avoidance or tolerance, during which the plant shoots grow faster under flooding conditions. This strategy increases dry matter production via photosynthesis when aerobic conditions are available above the water surface. *SNORKEL1* and *2* are involved in internode elongation, which enables flood avoidance and is a common strategy in floating rice. In this chapter, the mechanisms by which shoot elongation facilitates the avoidance of flood damage are presented.

5.2 Submergence Avoidance Strategy

Rice plants are thought to overcome flooding stress using one of two response strategies: submergence avoidance through the development of anatomical and morphological traits or submergence tolerance through the repression of cell elongation and carbohydrate metabolism (Bailey-Serres and Voesenek 2008). In general, rice plants, when exposed to low-oxygen conditions, such as the hypoxia that

J.-I. Sakagami (✉)
Faculty of Agriculture, Kagoshima University, Kagoshima, Japan
e-mail: Sakagami@agri.kagoshima-u.ac.jp

J.-I. Sakagami, M. Nakazono (eds.), *Responses of Plants to Soil Flooding*,
https://doi.org/10.1007/978-981-99-9112-9_5

occurs with flooding, show submergence avoidance. This involves the tendency to develop leaf blades and culms in an upward direction by accelerating the growth rate of the shoot when compared to normal growth conditions, to reach highly oxygenated aerobic conditions for gas exchange. This property is excellent for the avoidance of deep water. For example, floating and submerged rice areas where the arable land is flooded for long periods of time show remarkable internode and shoot elongation, and these characteristics are induced by rising water levels this helps facilitate submergence avoidance (Suge 1987). Floating and deep-water rice are often sown directly before the onset of flooding, and their growth is comparable to that of normal rice varieties during their early stages, but once flooding occurs, they grow taller, at a rate of approximately 8 cm per day (Kawano et al. 2008).

Floating and deep-water rice have adapted to prolonged flooding; the former mainly by increases in internode length and the latter by leaf elongation during the vegetative stages (Fig. 5.1). Floating rice is cultivated in flood-prone areas, and its mechanism of flooding avoidance by internode elongation may be influenced by a decrease in abscisic acid and an increase in gibberellin concentration or reactivity, in addition to an increase in the ethylene concentration in the internode spaces (Kende et al. 1998a, b).

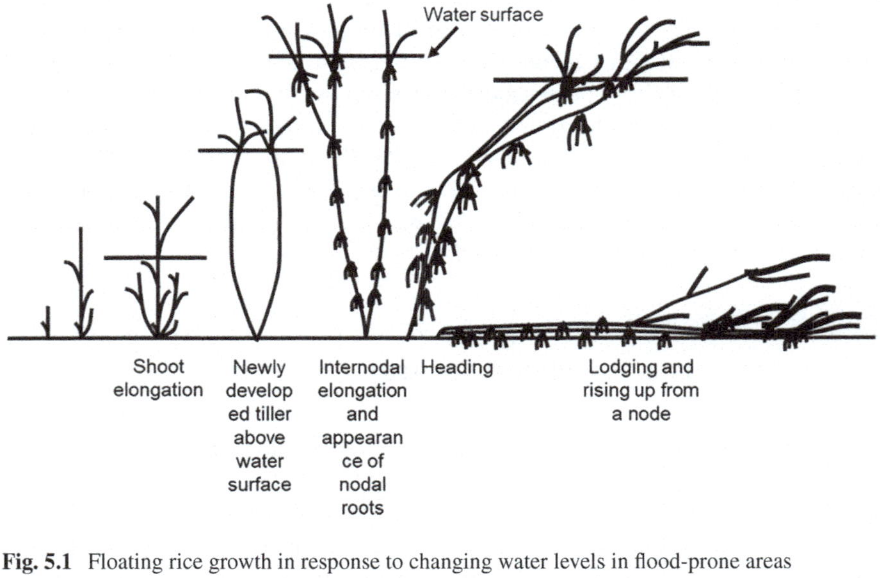

Fig. 5.1 Floating rice growth in response to changing water levels in flood-prone areas

5.3 Effects of Plant Hormones on Shoot Elongation

Various hormones are involved in the growth and physiological responses of plants. Ethylene is a specific plant hormone involved in fruit ripening, but importantly, it is also involved in rice growth. Ethylene generally promotes the lateral elongation of plants; however, for semi-aquatic and aquatic plants including rice, ethylene promotes vertical elongation (Ku et al. 1970). In the early stages of growth after seed germination in rice, the occurrence of coleoptile and mesocotyl elongation is most significant (Suge 1974). This is because ethylene promotes the translocation of sucrose from the scutellum to the mesocotyl, thereby increasing the accumulation of sugars in the mesocotyl, which contributes to elongation (Ishizawa and Esashi 1988). As gaseous diffusion in water is significantly reduced, gaseous ethylene accumulates in the rice plant body. Semi-aquatic plants such as rice have well-developed ventilated tissues, and once the leaf tips are in contact with the air, they can effectively capture atmospheric oxygen and transport it to the lower tissues. Rice plants can form air spaces in their root cortex called ventilation tissues, and aerenchyma can facilitate the diffusion of oxygen from the shoot to the root tips. The increased ethylene concentration in rice plants under water suggests that they may have acquired the ability to quickly elongate their shoots in response to ethylene to reach the aerobic conditions above the water surface and thus maintain their survival during their evolution.

In addition to ethylene, abscisic acid and gibberellins are closely related to submergence resistance. Internode elongation is an important trait in the avoidance of long-term flooding. When the water table rises due to flooding, the oxygen concentration in the internodes of rice plants decreases, and the carbon dioxide and ethylene concentrations increase. The increase in ethylene promotes a decrease in the abscisic acid concentration, which inhibits growth. This decrease in abscisic acid is due to the ethylene-induced expression of the gene encoding ABA 8′-hydroxylase, which converts abscisic acid to the less active phasic acid (Saika et al. 2007). Ethylene also increases the sensitivity to gibberellins, which act antagonistically to abscisic acid. This series of phytohormone actions results in the induction of internode elongation (Kende et al. 1998a, b).

5.4 *SNORKEL1* and *2* Genes Involved in Internode Elongation

In flood-prone areas, plants that can elongate their stems may avoid medium- to long-term submergence. Furthermore, in areas where plants thrive, stem elongation is closely related to environmental adaptation, as, for example, it can enable them to access lighter for photosynthesis. The SNORKEL1 (*SK1*) and SNORKEL2 (*SK2*) genes are closely related to internode elongation in floating rice plants (Hattori et al. 2009). Both genes were found to encode ethylene response factors involved in ethylene signalling, and they induce internode elongation through gibberellins.

Analyses have identified two factors that regulate gibberellin responsiveness in rice internodes. Subsequent analysis using molecular biological techniques led to the discovery of the gene ACCELERATOR OF INTERNODE ELONGATION1, which promotes internode elongation, and the gene DECELERATOR OF INTERNODE ELONGATION1, which inhibits it (Nagai et al. 2020). Thus, stem elongation is regulated and controlled by an intricate balance of accelerator and decelerator factors that control internode elongation, which is the mechanism that initiates stem elongation.

5.5 Photosynthetic Improvement as an Anaerobic Adaptation

In recent years, floods have caused increasing levels of crop damage, and specifically, flooding caused by torrential rains and overflowing rivers has inundated arable lands. This can deprive plants of the oxygen, carbon dioxide, and access to light that are required for growth. The resulting reduction in photosynthesis and respiration functions requires special plant growth features that can function under anaerobic conditions.

Rice leaves can reach above the water surface via shoot elongation when there is full submergence, thus enabling the resumption of aerobic metabolism and the fixation of carbon dioxide by photosynthesis. In general, the rate of gas exchange is slow in water (oxygen, 0.21 $cm^{-2}\ s^{-1}$ in air and $2.38 \times 10^{-5}\ cm^{-2}\ s^{-1}$ in water) due to the low diffusion coefficient of gases (Armstrong 1979). When water is stagnant, oxygen levels are even lower due the respiration of algae during the night. Rice increases alcohol fermentation under hypoxic conditions. However, alcohol fermentation produces only two molecules of ATP per glucose molecule, which is less efficient than aerobic respiration, which produces 32 molecules of ATP per glucose molecule. Consequently, rice plants cannot survive for long periods of time in hypoxic environments as they lack the carbohydrates required for energy production. Furthermore, photosynthesis is limited in water due to the low levels of solar radiation. To survive in water, it is thus also necessary to improve the photosynthetic capacity and effectively use photosynthetic products. For many rice varieties, avoidance of the anaerobic environment caused by flooding occurs through the development of leaf blades above the water surface, which then forms a canopy for photosynthesis (Setter et al. 1997; Ito et al. 1999; Kawano et al. 2002). A characteristic feature of the flooding avoidance genotype is the effective use of stored photosynthetic products for shoot elongation under hypoxic conditions when photosynthesis is severely restricted. However, failure to regain contact with oxygen, carbon dioxide, and light will inevitably lead to severe carbohydrate depletion.

The rapid increase in water depth accompanied by short-term flooding poses a high risk for young rice plants with a narrow leaf area, as it promotes above-ground elongation without dry matter production (Kawano et al. 2009). The results suggest that this rice species may be able to adapt to medium- to long-term flooding. This is because shoot elongation followed by active leaf unfolding improves photosynthesis and restores growth (Sakagami et al. 2013). For most rice varieties, rapid shoot elongation occurs during short periods of flooding; however, submergence-tolerant rice plants are able to adapt after flooding by suppressing shoot elongation in response to short periods of complete submergence (Fig. 5.2).

In general, when rice plants encounter flooding, they try to expand their leaf area above the water surface by altering the above-ground area. This strategy favors photosynthetic activity and improves dry matter production under anaerobic conditions. The photosynthetic rates of individual leaves from rice plants that were partially and fully submerged for 37 days were increased when compared with those that were not submerged (Sakagami et al. 2013). The photosynthetic rates with partial and complete submergence were closely related to the net assimilation rate during the submergence period. This submergence response may facilitate the translocation of photosynthetic assimilates to the internodes and growth organs of rapidly growing culms (Raskin and Kende 1984). The phenomenon of increased photosynthetic rates under submerged conditions may be a means of increasing the activity of leaves that can survive under anaerobic conditions, resulting in higher dry matter weight per unit area.

During short periods of complete submergence, rapid shoot elongation is likely to result in reduced growth and decline after water withdrawal, and the survival strategy of submergence-tolerant rice plants carrying the SUB1A gene is to suppress shoot elongation during flooding to conserve energy and efficiently consume photosynthetically assimilated products that were assimilated before submergence. This strategy also prepares the plant for the environmental changes after

Fig. 5.2 Comparison of susceptible and tolerant rice varieties 7 days after complete submergence and 7 days after desubmergence during the recovery period. (Modified from Sakagami et al. (2010))

submergence. In addition, the translocation rate of assimilated products by photosynthetic during submergence is faster in susceptible varieties than in tolerant varieties. It is highly likely that this characteristic will influence energy use efficiency (Izzawat et al. 2022).

5.6 Adaptation to Prolonged Flooding and Deep Water in Specific Rice Genotypes

Globally, there are two predominant and distinct rice species: Asian rice (*O. sativa* L.) and African rice (*O. glaberrima* Steud). Each has different characteristics, as the Asian rice genotype varies from annual to perennial, whereas the African rice genotype is generally a monocarpic annual. In addition, African rice is thought to be cultivated in the Niger River delta region of Mali, which suggests that it is highly adapted to prolonged flooding. Consequently, the effects of short-term complete submergence on the early growth of these different species were compared. The African rice varieties showed greater shoot elongation ability when compared with the Asian rice varieties under complete submergence conditions with a 45 cm water depth in a paddy field for 10 days. However, the African rice varieties were found to lodge easily after desubmergence due to its elongated shoots, and this decreased its survival rate. It was concluded that this was due to the loss of photosynthetic carbohydrates and cell elongation caused by the short period of rapid shoot elongation, which did not respond to the different late-stage conditions after desubmergence. African rice exhibits rapid growth characteristics with shoot elongation upon complete submergence, which is an important feature in flood-prone areas (Joho et al. 2008). Luo et al. (2020) found that the submergence tolerance of *O. glaberrima* is elicited through a different mechanism than that of the tolerant *O. sativa* cultivar. An experiment was then conducted with 57 cultivars, 30 cultivars of *O. sativa* and 27 of *O. glaberrima*, in a paddy field in the Republic of Guinea (Fig. 5.3, Table 5.1). Twenty days after transplanting, all plants were completely submerged under 40 cm of water for 30 d. The average shoot elongation rate (cm d^{-1}) and shoot dry matter weight increases (g $plant^{-1}$ d^{-1}) during the submergence were 1.63 and 1.97 and 0.04 and 0.10 for *O. sativa* and *O. glaberrima*, respectively. Furthermore, most of the *O. glaberrima* cultivars demonstrated a greater increase in tiller number during submergence and post-submergence recovery than the *O. sativa* cultivars.

Fig. 5.3 Growth of African rice variety *O. glaberrima* when completely submerged in Guinea. Flooding treatments were performed 14 days after transplanting for 30 days

5.7 Rice Adaptation to Flood-Prone Areas in Africa

The Niger River, which crosses West Africa and is constantly flowing throughout the year, is a valuable source of water in this region for industry, agricultural production, and people's livelihoods. Along the Niger River, from Mopti to Timbuktu in Mali, an inland delta is created by the overflow of the main river into its tributaries, and this forms a large floodplain during the rainy season (Fig. 5.4). The rainy season in the Guinea Plateau region upstream of the Niger River begins around June, and the flow rate in the inland delta then increases rapidly from July to October, and river water flows out and is stored in the floodplains. This is considered advantageous for rice development in arid lands as there is a reliable water source. In general, the soil in this area is more fertile than that in the upland fields near the floodplain, due to the input of manure from livestock and the supply of nutrients from the flood waters. African rice is grown exclusively in floodplain watersheds. This is because it is considered suitable for different water stress environments: soil drying in the early growing season and flooding from the middle of the growing season onward. Sowing is performed at the beginning of the rainy season, and harvesting takes place around December during the dry season, regardless of the variety. If the soil is wet at the time of sowing, the sowing is done first and then cultivated to maintain healthy germination. However, if the soil is dry, the seeding is done after plowing and the first rainfall. The water levels in the floodplain can increase and decrease the topography changes. A few weeks after seeding, the inflow of floodwater causes the rice plants to grow rapidly, and in some places, the water level rises to more than 2 m during the reproductive period. As the water table rises, the rice plants repeatedly elongate their internodes, which also produce nodal roots. Although the function of nodal roots has not been fully clarified, they may facilitate

Table 5.1 Shoot elongation rate and shoot biomass of *O. sativa* and *O. glaberrima* show different rates of increase under complete submergence

		Shoot length (cm)			Shoot biomass (g plant^{-1})		
Treatments	Species	*A*	*B*	Shoot elongation (cm d^{-1})	*A*	*B*	Shoot biomass increase (g d^{-1})
Non-submergence	*O. sativa* L. (*n* = 30)	28.4 ± 0.85	65.6 ± 1.48	1.20 ± 0.04	0.16 ± 0.01	12.2 ± 0.49	0.39 ± 0.02
	O. glaberrima Steud. (*n* = 27)	31.5 ± 0.94	64.9 ± 1.47	1.08 ± 0.05	0.17 ± 0.01	11.7 ± 0.56	0.37 ± 0.02
T-test (*O. sativa* × *O. glaberrima*)		*	NS	NS	NS	NS	NS
Submergence	*O. sativa* L. (*n* = 30)	28.2 ± 0.67	78.8 ± 1.99	0.63 ± 0.05	0.11 ± 0.005	1.32 ± 0.18	0.04 ± 0.01
	O. glaberrima Steud. (*n* = 27)	31.0 ± 0.95	91.9 ± 1.65	1.97 ± 0.03	0.12 ± 0.004	3.38 ± 0.22	0.10 ± 0.01
T-test (*O. sativa* × *O. glaberrima*)		*	**	**	**	**	**

A, *B*: before, after submergence. Source: Jun-Ichi et al. (2009)
NS indicates the non significant differences between spiecies, determined by t-test
* and ** indicate significant difference ($p<0.05$ and $p<0.01$)

Fig. 5.4 Cultivation of the *O. glaberrima* rice cultivar in the flood-prone area of Niger river. 2004, in Mopti of Republic of Mali

the absorption of nutrients that are contained in the floodwater. In addition, elongated rice plants rapidly develop leaf blades on the water surface, allowing for biomass production through photosynthesis and the maintenance of aerobic respiration. Furthermore, the rising water table inhibits weed seed germination and reproduction, and consequently, farmers do not weed at all during the growing season. Harvesting is done by panicle cutting if the water is deep enough and, in some cases, even in small boats. Thus, African rice is an indispensable rice species in the region, contributing to the maintenance of a sustainable, recycling-oriented cultivation system that takes advantage of the deep waters that occur in the floodplains.

References

Armstrong W (1979) Aeration in higher plants. Adv Bot Res 7:225–331

Bailey-Serres J. Voesenek LACJ (2008) Flooding stress: acclimations and genetic diversity. Annu Rev Plant Biol 59:313–339

Hattori Y, Nagai K, Furukawa S, Song XJ, Kawano R, Sakakibara H, Wu J, Matsumoto T, Yoshimura A, Kitano H, Matsuoka M, Mori H, Ashikari M (2009) The ethylene response factors SNORKEL1 and SNORKEL2 allow rice to adapt deep water. Nature 460:1026–1030

Ishizawa K, Esashi Y (1988) Action mechanism of ethylene in the control of sugar translocation in relation to rice coleoptile growth. I Sucrose metabolism. Plant Cell Environ 29:131–141

Ito O, Ella E, Kawano N (1999) Physiological basis of submergence tolerance in rainfed lowland rice ecosystem. Field Crops Res 64:75–90

Izzawati NAH, Yabuta S, Junaedi A, Sakagami J-I (2022) Different survival strategies involve carbon translocation rather than de novo C assimilation under complete submergece in rice plant. Photosynthesis Reseach 154:183–193

Joho Y, Omsasa K, Kawano N, Sakagami J-I (2008) Growth responses of seedlings in Oryza glaberrima Steud. to short-term submergence in Guinea, West Africa. Jpn Agric Res Quar 42:157–162

Jun-Ichi S, Joho Y, Ito O (2009) Contrasting physiological responsees by cultivars of *Oryza sativa* and *O. glaberrima* to prolonged submergece. Ann Bot 103:171–180

Kawano N, Ella E, Ito O, Yamauchi Y, Tanaka K (2002) Metabolic changes in rice seedlings with different submergence tolerance after desubmergence. Environ Exp Bot 47:157–203

Kawano N, Ito O, Sakagami J-I (2008) Flash flooding resistance of rice genotypes of Oryza sativa L., *O. glaberrima* Steud. and Interspecific hybridization progeny. Environ Exp Bot 63:9–18

Kawano N, Ito O, Sakagami J-I (2009) Morphological and physiological responses of rice seedlings to complete submergence (flash flooding). Ann Bot 103:161–169

Kende H, Van der Knaap E, Cho H-T (1998a) Deepwater rice: A model plant to study stem elongation coleoptile growth in internodes of deepwater rice. Planta 90:333–339

Kende H, van der Knaap E, Cho HT (1998b) Deepwater rice: A model plant to study stem elongation. Plant Physiol 118:1105–1110

Ku HS, Suge H, Rapport L, Pratt HK (1970) Stimulation of rice coleoptile growth by ethylene. Planta 90:333–339

Luo Q, Nakazawa M, Sasayama D, Fukayama H, Hatanaka T, Azuma T (2020) Responses to complete submergence in African Rice (*Oryza glaberrima*). J Crop Res 65:37–41

Nagai K, Mori Y, Ishikawa S et al (2020) Antagonistic regulation of the gibberellic acid response during stem growth in rice. Nature 584:109–114

Raskin I, Kende H (1984) Effect of submergence on translocation, starch content and amylolytic activity in deep-water rice. Planta 162:556–559

Saika H, Okamoto M, Miyoshi K, Kushiro T, Shinoda S, Jikumaru Y, Nakazono M (2007) Ethylene promotes submergence-induced expression of OsABA8ox1, a gene that encodes ABA 8′-hydroxylase in rice. Plant Cell Physiol 48:287–298

Sakagami J-I, Mochizuki T, Watanabe H (2010) Wet land and crop production. In: Sakagami J-I, Nakazono M, Shimamura S, Ito O, Ishizawa K (eds) Surviving strategy and submergence resistance in rice plants. Yokendo, Tokyo

Sakagami J-I, Joho Y, Sone C (2013) Complete submergence escape with shoot elongation ability by underwater photosynthesis in African rice, *Oryza glaberrima* Steud. Field Crop Res 152:17–26

Setter TL, Ellis M, Laureles EV, Ella ES, Senadhira D, Mishra SB, Sarkarung S, Datta S (1997) Physiology and genetics of submergence tolerance in rice. Ann Bot 79:67–77

Suge H (1974) Synergistic action of ethylene with gibberellins in the growth of rice seedlings. Proc Crop Sci Soc Jpn 43:83–87

Suge H (1987) Physiological genetics of internodal elongation under submergence in floating rice. Jpn J Genet 62:69–80

Chapter 6
The Importance of Leaf Gas Films for Gas Exchange During Submergence

Ole Pedersen and Max Herzog

6.1 Water-Repellent Leaves

Superhydrophobic leaves are found in several families of both dicots and monocots, and hydrophobicity is thought to have evolved as a self-cleansing feature. In a comprehensive study of 200 plant species, the surface structures resulting in hydrophobicity were meticulously examined and then classified using SEM (Neinhuis and Barthlott 1997). The most prevailing nano-structures resulting in hydrophobicity were platelets and tubules, but other structures such as rodlets, ribbons, and bands were observed as well. Importantly, the 200 species examined all had contact angles of micro droplets exceeding 150° (Neinhuis and Barthlott 1997), which is used to separate hydrophobic from superhydrophobic (Koch and Barthlott 2009). Few, if any, of the examined species were naturally exposed to flooding and superhydrophobicity, therefore considered a self-cleansing mechanism to prevent that dirt and pathogens (spores from bacteria and fungi) accumulate on the leaf surfaces, as rolling water droplets naturally pick up and remove such particles during rain (Barthlott et al. 2017).

Hydrophobicity derives from hierarchical structures on the leaf surfaces. In the case of *Nelumbo nucifera* (sacred lotus), where leaf hydrophobicity has been most thoroughly studied, hydrophobicity is the result of micro (papillae)- and nano (wax platelets)-structures (Bhushan et al. 2009). The stunning hydrophobicity of sacred lotus has given rise to the "lotus effect" referring to superhydrophobic surfaces

O. Pedersen (✉)
Department of Biology, University of Copenhagen, Copenhagen, Denmark

School of Biological Sciences, The University of Western Australia, Crawley, WA, Australia
e-mail: opedersen@bio.ku.dk

M. Herzog
Department of Biology, University of Copenhagen, Copenhagen, Denmark

J.-I. Sakagami, M. Nakazono (eds.), *Responses of Plants to Soil Flooding*,
https://doi.org/10.1007/978-981-99-9112-9_6

(Spaeth and Barthlott 2008). In rice, an additional dimension is added, i.e., the macro-structures deriving from the plicate leaves (Wu et al. 2011) so that superhydrophobicity of rice leaves is the combined effect of the grooves formed by the plicate leaves (macro-structures), the papillae covering the leaves (micro-structure), and the wax platelets forming the surface of the cuticle (nano-structures).

The heritability of leaf hydrophobicity is not known in detail, but the best studied level is the nano-structures. Several rice mutants are known to show contrasting phenotypes in terms of the density of wax platelets (Kurokawa et al. 2018; Zhang et al. 2016), and expression of *LGF1* (*OsHSD1*) is required for maximum hydrophobicity. The are also mutants with contrasting densities of papillae (the microstructure) (Yoo et al. 2011), and *Oryza* species are also known to show contrast in the degree of plicate leaves (Chatterjee et al. 2016). Consequently, the relative importance of macro-, micro- and nano-structures for leaf hydrophobicity could be experimentally evaluated for rice.

6.2 Gas Film Retention of Superhydrophobic Leaves Under Water

Superhydrophobic leaves retain a thin gas film when submerged into water. In some species, the hydrophobicity is so strong that even if the leaf has never been in contact with air (i.e., when it emerges from an already submerged stem), a thin gas film is nevertheless formed on these leaves (Fig. 6.1). The mechanism behind the

Fig. 6.1 Leaf gas films on completely submerged plants. Some wetland plants possess superhydrophobic leaves where gas films are formed on surfaces that have never been in contact with the atmosphere. (**a**) shows the young emergent shoots of *Phragmites australis* (common reed) at a depth of about 1 m; the gas films are visible as a silvery sheen on the pointy shoots. In (**b**), the leaf of *Spartina anglica* (common cordgrass) is completely submerged in saline water, where algae have covered the older decaying part of the shoot. In contrast, the young superhydrophobic leaf is clean since the algae cannot attach to the gas layer. Photos by Ole Pedersen

formation of gas films on surfaces that have not been in contact with the atmosphere has not been studied, but a sudden increase in water temperature causing the solubility of gasses to exceed air equilibrium could possibly initiate the gas film formation. Moreover, underwater photosynthesis producing O_2 could also act as the spark igniting gas film formation on the superhydrophobic leaf cuticles.

Gas film formation on submerged leaves was first reported for deepwater rice, wheat, barley, and oats, where the beneficial effects on carbon fixation were also first reported (Raskin and Kende 1983; see below). Later, a series of studies reported leaf gas films in several species of submerged wild wetland plants, where gas films were retained on partially or completely submerged leaves (Fig. 6.2).

The gas films on plicate leaves are mainly retained in the deep grooves between each parallel ridge. The 3D distribution of gas layers on leaves of *Spartina anglica* has been visualized using X-ray micro tomography, and it showed that the vast majority of gas volume was positioned in between the ridges with only a mesh-like gas structure connecting the longitudinal bodies of gas (Lauridsen et al. 2014). However, this is the only study that has revealed the 3D structure of the gas films, and all other studies are simply reporting a mean gas film thickness using a buoyancy approach to assess the total volume of gas and the area of the leaf surface to calculate thickness. Using this quantitative approach, the gas film thickness of seven species of monocots and only one dicot species (*Brassica napus*) has been evaluated immediately after submergence. Here, we have summarized the gas film thicknesses from the literature showing that leaf gas film thickness of rice or wheat does not

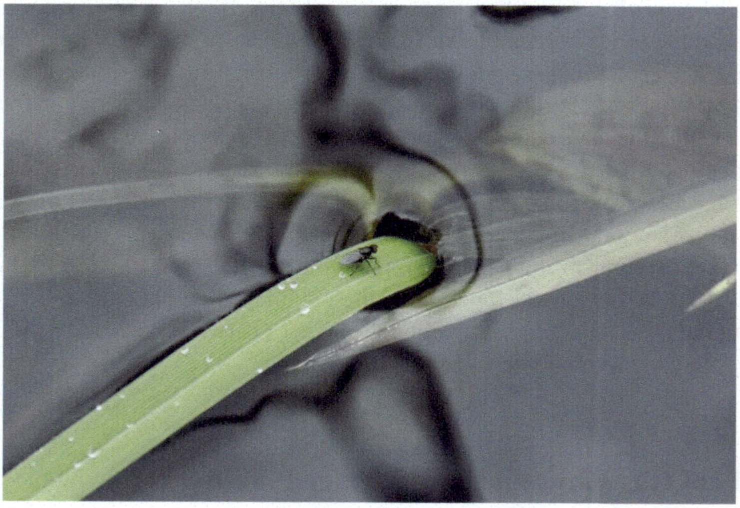

Fig. 6.2 Superhydrophobic leaves of *Glyceria fluitans*. The plicate leaf surfaces are coated by microscopic wax platelets resulting in water repellence. The water droplets on the floating leaf forms ideal tiny spheres with contact angles exceeding 150°, and the fly is looking into the cavity that continues under water coating the submerged leaves with a thin gas film (visible as the silvery sheen). Photo by Ole Pedersen

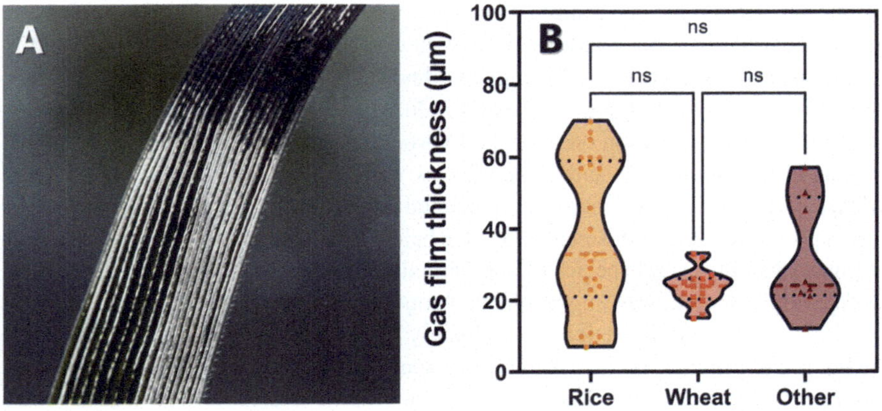

Fig. 6.3 Leaf gas films form on submerged superhydrophobic leaves in several plant species. Like other cereals, submerged leaves of wheat retain a gas film visible as a silvery sheen (**a**). Leaf gas film thickness varies significantly among species, and (**b**) shows the thickness in rice ($n = 25$), wheat ($n = 16$), and other species (*Avena sativa*, *Brassica napus*, *Glyceria fluitans*, *Hordeum vulgare*, and *Spartina anglica*; $n = 8$) immediately after submergence before the leaf hydrophobicity starts declining. Photo in (**a**) by Ole Pedersen. Literature data in (**b**) from (Chakraborty et al. 2021; Herzog et al. 2018b; Konnerup and Pedersen 2017; Konnerup et al. 2017; Kurokawa et al. 2018; Lauridsen et al. 2014; Mori et al. 2019; Pedersen et al. 2009; Raskin and Kende 1983; Verboven et al. 2014; Winkel et al. 2011, 2013, 2014, 2017; Wittig et al. 2021). Data were tested using a non-parametric Kruskal-Wallis test

differ from that of other species tested (*Avena sativa*, *Brassica napus*, *Glyceria fluitans*, *Hordeum vulgare*, and *Spartina anglica*) (Fig. 6.3).

An example of leaves combining superhydrophobicity with hydrophilic properties is the fronds of *Salvinia molesta*, a floating tropical fern. The fronds are densely covered by numerous hairlike structures, and hairs are grouped together in four with the ends facing the environment (Barthlott et al. 2010). In mature leaves, the terminal cell of each hair is dead forming a cap. The whole leaf is covered with wax crystals except for the four terminal cells, and as a result, the surface of the terminal cells is rather smooth, whereas the wax crystals create a nanoscale roughness on all other cells, resulting in the superhydrophobic properties of *S. molesta* (Barthlott et al. 2010). The ecological function of the complexed architecture is not totally clear, but the fern can retain large volumes of gasses between the hairs, helping it to surface following a submergence event.

6.3 Leaf Gas Films and Underwater Photosynthesis

The influx of CO_2 from floodwater to leaf tissues is substantially enhanced by leaf gas films, thereby stimulating underwater net photosynthesis. Enhanced CO_2 fixation in the presence of leaf gas films was first shown in a pioneering study on deepwater rice using radio-labelled inorganic carbon (^{14}C) (Raskin and Kende 1983). This study revealed a tenfold higher CO_2 fixation rate in control leaves with gas films as compared to leaves where the gas films had been experimentally removed by using a dilute detergent to reduce hydrophobicity. Although CO_2 fixation was only tested on rice, Raskin and Kende (1983) showed that other cereals, i.e., wheat, barley, and oats, also retain a gas film when submerged into water, but the effect on underwater photosynthesis was not assessed.

The wider importance of superhydrophobic leaves for underwater net photosynthesis was shown in a multi-species study on wild wetland plants. Here, the four species with superhydrophobic leaves all retained a gas film when submerged, which significantly enhanced underwater net photosynthesis (Colmer and Pedersen 2008). The relative importance of leaf gas films was higher at 500 μM CO_2 in the floodwater compared to only 50 μM CO_2, and on average, the leaf gas films resulted in three- to sevenfold higher rates in net photosynthesis. Interestingly, a CO_2 response curve for *Phragmites australis* showed a characteristic saturation response similar to that observed for leaves in air, but underwater net photosynthesis was only saturated at around 1500–2000 μM CO_2, which is more than 100-fold that of air equilibrium. The latter clearly demonstrates that the CO_2 influx is still diffusion limited under water even in the presence of leaf gas films.

Wheat, like other cereals, also possesses superhydrophobic leaves, and therefore these retain a gas film during submergence. Submergence of wheat can happen particularly in the early seedling stages, and during such events, the underwater net photosynthesis of wheat also benefits from leaf gas films. In the presence of gas films, underwater net photosynthesis becomes CO_2 saturated at around 1000 μM, and the rates are about sixfold higher in leaves with gas films compared to leaves without (Winkel et al. 2017). Interestingly, completely submerged wheat seedlings survived about 3 days longer with intact gas films as compared to submerged wheat where the gas films had been experimentally removed by reducing leaf hydrophobicity (Winkel et al. 2017). However, in a study of two genotypes of wheat showing contrasting tolerance to submergence, the differences in survival could not be related to differences in gas film retention time (Herzog et al. 2018a).

In summary, leaf gas films forming on superhydrophobic leaves of terrestrial wetland plants significantly enhance underwater photosynthesis. The gas films facilitate the flux of CO_2, which is dissolved in the floodwater, into leaves, and O_2 produced in photosynthesis can also escape the photosynthetic tissues via the leaf gas films. At environmentally relevant CO_2 concentrations, underwater net photosynthesis of leaves with gas films is six- to tenfold higher than in leaves where the hydrophobicity has been removed. Nevertheless, the photosynthetic rates obtained under water are typically only 10% of maximum rates obtained in air, demonstrating that the slow diffusion of gasses in water still restricts gas exchange even in the presence of leaf gas films.

6.4 Leaf Gas Films and Underwater Respiration

During darkness, the O_2 flux from floodwater to the leaf tissues is also greatly facilitated by leaf gas films. A multi-species study on wild wetland plants showed that the critical O_2 pressure for underwater respiration ranged from 60 to 130 μM O_2 (pO_2 = 4.4–9.4 kPa) for leaves of plants with superhydrophobic leaves forming a gas film, whereas for hydrophilic leaves, the critical O_2 pressure exceeded air equilibrium and could not be determined using the applied experimental approach (Colmer and Pedersen 2008). This study also manipulated the leaf hydrophobicity of *P. australis* to prevent gas film formation, and these experiments revealed a typical saturation curve for O_2 consumption vs. external O_2 concentration in control leaves with gas films that saturated at around 60 μM O_2 (pO_2 = 4.4 kPa). In contrast, leaves without a gas film did not show sign of saturation in medium containing O_2 at air equilibrium (284 μmol O_2 or 20.6 kPa) (Colmer and Pedersen 2008).

Interestingly, O_2 exchange with water is also facilitated by thin gas films in some aquatic insects and spiders. Superhydrophobic gas-retaining cuticle surfaces relevant for respiration can be classified into two categories, (i) "physical gills," also referred to as compressible gas gills, and (ii) true plastrons, also referred to as incompressible gas gills (Balmert et al. 2011). The plastron is similar to the leaf gas films in plants, and leaf gas films have actually been referred to as plant plastrons (Raven 2008). Due to the physical function of plastrons, there is a striking similarity between plants and insects when analyzing the relationship between O_2 consumption (respiration) and dissolved O_2 in the water (Pedersen and Colmer 2012). The example of *Phalaris arundinacea* (wetland plant) and *Aphelocheirus aestivalis* (aquatic insect) has an almost identical response with the only difference being that the maximum consumption rate of the insect is about twice that of the leaves when expressed per surface area; this is likely due to the much higher protein content of the insect. Apart from this difference, the critical O_2 pressure for respiration is similar, i.e., around 100 μM O_2 (pO_2 = 7.2 kPa), and the effect of removing the plastron is also similar (Fig. 6.4). Consequently, these physical gills prevent drowning of many wetland insects, spiders, and plants.

The beneficial effects of leaf gas films on O_2 status are not only restricted to aboveground tissues. A study using O_2 microsensors to monitor cortical O_2 status in roots of submerged rice demonstrated the effect of leaf gas films on internal aeration in light as well as in darkness (Pedersen et al. 2009).

6.5 Gas Film Retention Time During Submergence

Leaf hydrophobicity declines with time of submergence resulting in loss of gas films. Four genotypes of rice that were completely submerged lost their leaf gas films within 4–7 days of submergence (Winkel et al. 2014). Following loss of gas films, underwater net photosynthesis declined, and chlorophyll senescence

Fig. 6.4 Leaf gas films on wetland plants and plastrons of aquatic insects facilitate O_2 exchange with the surrounding water. The leaf gas films are visible as a silvery sheen on the submerged leaves (**a**), and the plastron can also be seen as a silvery layer of gas (**c**). O_2 consumption in both plant and insect shows striking similarities, and both organisms respond in the same way to removal of hydrophobicity by brushing the surface with a dilute detergent to prevent formation of gas films or plastron (**b**, **d**); after removal of hydrophobicity, the O_2 consumption becomes diffusion limited across the entire range of external O_2 tested. Figure modified from Pedersen and Colmer (2012)

accelerated, but gas film retention time was not related to the *SUB1*, which was represented in two of the tested genotypes (Winkel et al. 2014). Meanwhile, a later study found that leaf gas film thickness was higher in *SUB1*-genotypes than genotypes not containing the *SUB1* locus, which was related to lower expression of *LGF1*-gene in the latter (Chakraborty et al. 2021). For rice submerged in a field situation, the benefits of superhydrophobic leaves and the associated leaf gas films as related to underwater photosynthesis and respiration are restricted to about 1 week of complete submergence, whereas under controlled laboratory conditions with submergence into clean water, the gas films can persist longer (Pedersen et al. 2009).

Young seedlings of wheat are also prone to complete submergence in some environments. Wheat, as well as other cereals (Raskin and Kende 1983), also possesses superhydrophobic leaves, and therefore these retain a gas film when submerged. In a study including several genotypes, gas film retention of wheat during submergence was tested, but this study found no differences in gas film retention time among the 14 genotypes tested (Konnerup et al. 2017). However, 13 out of 14 tested wheat cultivars were Danish cultivars, and genetic diversity would therefore be limited compared to a global subset of cultivars. Moreover, the hydrophobicity (and thereby also the gas films) was lost within only 2 days of submergence with scanning electron microscopy, suggesting that the loss of hydrophobicity was caused by a layer of dirt or biofilm covering the nano-structures (wax platelets) of the leaves (Konnerup et al. 2017). Interestingly, an experimental study manipulating the leaf hydrophobicity showed that the better survival of completely submerged wheat, which initially had superhydrophobic leaves, matched the gas film retention time, i.e., leaf gas films were lost after 2 days of submergence, and the plants starting out with leaf gas films survived exactly 2 days longer (Winkel et al. 2017).

Gas film retention has also been studied in young seedlings of *Brassica napus*. Two genotypes were tested with one of them retaining a gas film when submerged. The genotype, Avatar, retaining leaf gas films upon submergence also had leaves with contact angles exceeding 150 ° (Wittig et al. 2021), rendering it superhydrophobic (Koch and Barthlott 2009). However, the superhydrophobic properties were lost within 24 h after submergence, which is possibly the reason why there was no difference in survival of the two genotypes following 3 weeks of complete submergence (Wittig et al. 2021).

6.6 The Effect of Leaf Gas Films Under Field Conditions

The beneficial effect of leaf gas films for underwater gas exchange has been demonstrated in the field situation for *Spartina anglica*, which is exposed to tidal submergence. In some coastal wetlands, *S. anglica* (common cordgrass) becomes completely submerged twice every day during high tides (Winkel et al. 2011). Isolated clones of plants, which had their leaves brushed with a dilute detergent to reduce hydrophobicity and thereby prevent gas film formation during high tides, had significantly lower tissue O_2 status in the belowground organs both when submerged in darkness and in the light (Winkel et al. 2011). This clearly shows that leaf gas films facilitate CO_2 supply from the tidal water to the leaves during the light and O_2 during darkness. Interestingly, the leaves of *S. anglica* are only superhydrophobic on the adaxial side, and this is exactly where the stomata are positioned. In contrast, the abaxial side is hydrophilic, and there are no stomata in the epidermis. Therefore, this side of the leaf becomes dirty as mud particles in the tidal water stick to the hydrophilic cuticle (Winkel et al. 2011), showing the dual purpose of superhydrophobicity, i.e., facilitate gas exchange during submergence and act as a self-cleansing mechanisms preventing particles to adhere to the leaf cuticle.

During complete submergence, leaf gas films also enhance underwater gas exchange in paddy rice. Particularly the vegetative stage of paddy rice is prone to submergence, and a study conducted in a real paddy situation demonstrated the beneficial effects of leaf gas films during complete submergence (Winkel et al. 2013). The study used an experimental approached and removed hydrophobicity from one set of plants with another set of plants serving as controls; the controls retained a gas film during submergence, whereas the plants that had been brushed with a dilute detergent did not. Oxygen microsensors were positioned in the roots near the root-shoot junction, and the O_2 measurements revealed that the O_2 status of roots of rice with leaf gas films were significantly better during submerged, and the beneficial effect remained substantial during the day as well as during the night (Winkel et al. 2013). This study also demonstrated the linear relationship between flood water O_2 content and root tissue O_2 status during the night. During the day, light insolation determined root O_2 status in completely submerged plants with only a minor contribution from the floodwater, i.e., the O_2 inside the roots was derived from underwater photosynthesis (Winkel et al. 2013).

Leaf gas films have been studied in a wide range of wild plants growing in a natural flooding gradient. A study evaluated if species growing in habitats that were prone to flooding were also more likely to have leaf gas films as compared to species growing in habitats that were less prone to flooding. Interestingly, the study found the exact opposite relationship between number of species with gas films and the flooding risk; the lower the flooding risk, the higher the frequency of plants with leaf gas films (Winkel et al. 2016). The study concluded that the pattern was driven by grass species that all have superhydrophobic leaves; grasses simply became more frequent as the flooding risk declined (Winkel et al. 2016). However, the study also concluded that the specific leaf area was positively related to flooding risk, showing the importance of underwater photosynthesis during flooding since thin leaves (high specific leaf area) photosynthesize better under water than thick leaves (low specific leaf area) (Colmer et al. 2011).

The beneficial effects of hydrophobicity of aquatic plants are not always related to performance during complete submergence. The floating leaves of *Glyceria fluitans* are only superhydrophobic on the adaxial side of the leaves facing the atmosphere, whereas the abaxial side becomes hydrophilic when in contact with water (Konnerup and Pedersen 2017). *G. fluitans* nevertheless showed high rates of underwater photosynthesis, and these were obtained at environmentally relevant CO_2 concentrations, but it appeared that the floating leaves possess both aquatic and terrestrial properties and thus have "the best of both worlds" so that floating leaves are particularly adapted to situations where the plant is partially submerged and occasionally experiences complete submergence (Konnerup and Pedersen 2017).

6.7 Gas Films Can Restrict Salt Intrusion During Saline Submergence

Leaf gas films can act as a physical barrier between the leaf surface and the surrounding floodwater so that intrusion of salt is restricted during submergence in saline water. The insulating properties of gas films between a saline environment and the leaf surfaces have been demonstrated for *Melilotus siculus*, which is a salinity and waterlogging tolerant legume (Teakle et al. 2012). The study demonstrated that leaf gas films significantly reduced Na^+ and Cl^- intrusion into leaves when submerged in saline water—the thin gas layer physically separated the floodwater from the leaf surface. This feature aids survival of plants exposed to short-term saline submergence, as well as the previously recognized beneficial effects of gas exchange under water (Teakle et al. 2014).

Rice can also become submerged by saline water; therefore, the effect of leaf gas films on salt intrusion was tested in a laboratory experiments. Leaves of rice were brushed with a dilute detergent to remove hydrophobicity and thereby prevent gas film formation during submergence and compared to control plants with superhydrophobic leaves (Herzog et al. 2018b). The study found that in spite of leaf gas film presence, Na^+ entered the rice leaf, albeit at a slower rate than in leaves with the gas film removed when completely submerged in saline water. In contrast, there was no delay on Cl^- intrusion (Herzog et al. 2018b). The insulating effect of the gas films was less than anticipated, indicating the presence of some leaf-to-floodwater contact, so that the gas layer must not have completely separated the leaf surfaces from the water. Such a leaf-to-floodwater contact could have been caused by the mesh-like structure of the gas film observed for *S. anglica* (Lauridsen et al. 2014) (see above).

6.8 Conclusion and Outlook

Since the pioneering work on deepwater rice (Raskin and Kende 1983), a great body of literature has demonstrated the beneficial effects of superhydrophobic leaves during submergence of terrestrial plants. The retention of a thin gas films on the cuticle of superhydrophobic leaves significantly enhances gas exchange with the floodwater, thereby partly overcoming the slow diffusion of gasses in water compared to in air. Leaf gas films thereby enhance underwater photosynthesis by increasing the flux of CO_2 into the submerged leaf, and in darkness, the gas films increase the flux of O_2 so that dark respiration is sustained. These effects are not only restricted to the aboveground tissues posing gas films but is also extended to belowground tissues as demonstrated by much higher O_2 status in belowground tissues in the presence of gas films. However, leaf hydrophobicity—and thereby the gas films—is lost within a few days of submergence.

At present, the reason behind the loss of hydrophobicity during submergence is unknown. Since the beneficial effects are undisputable, future research should focus on unravelling the causal effects behind the declining hydrophobicity. We therefore propose to pursue a novel approach using artificial surfaces where key parameters such as depth and angles of the grooves of artificial leaves can be manipulated following the approach of Lu et al. (2017). Moreover, a recent study has demonstrated that it is now possible to mimic the nano-coating of wax platelets, which is a feature of superhydrophobic leaves, using artificial materials (Huth et al. 2022). We predict that combining these two approaches into a single study holds the potential for greatly advancing our understanding of loss of hydrophobicity under water, and this knowledge could be utilized to breed more flood-tolerant crops pending that the genetics behind the features leading to hydrophobicity in plants are also known.

References

Balmert A, Florian Bohn H, Ditsche-Kuru P, Barthlott W (2011) Dry under water: comparative morphology and functional aspects of air-retaining insect surfaces. J Morphol 272:442–451. https://doi.org/10.1002/jmor.10921

Barthlott W, Schimmel T, Wiersch S, Koch K, Brede M, Barczewski M, Walheim S, Weis A, Kaltenmaier A, Leder A, Bohn HF (2010) The *Salvinia* paradox: superhydrophobic surfaces with hydrophillic pins for air retention under water. Adv Mater 22:2325–2328. https://doi.org/10.1002/adma.200904411

Barthlott W, Mail M, Bhushan B, Koch K (2017) Plant surfaces: structures and functions for biomimetic innovations. Nano-Micro Lett 9:23. https://doi.org/10.1007/s40820-016-0125-1

Bhushan B, Jung YC, Koch K (2009) Micro-, nano-and hierarchical structures for superhydrophobicity, self-cleaning and low adhesion. Philos Trans R Soc A Math Phys Eng Sci 367:1631–1672. https://doi.org/10.1098/rsta.2009.0014

Chakraborty K, Guru A, Jena P, Ray S, Guhey A, Chattopadhyay K, Sarkar RK (2021) Rice with SUB1 QTL possesses greater initial leaf gas film thickness leading to delayed perception of submergence stress. Ann Bot 127:251–265. https://doi.org/10.1093/aob/mcaa171

Chatterjee J, Dionora J, Elmido-Mabilangan A, Wanchana S, Thakur V, Bandyopadhyay A, Brar DS, Quick WP (2016) The evolutionary basis of naturally diverse rice leaves anatomy. PLoS One 11:e0164532. https://doi.org/10.1371/journal.pone.0164532

Colmer TD, Pedersen O (2008) Underwater photosynthesis and respiration in leaves of submerged wetland plants: gas films improve CO_2 and O_2 exchange. New Phytol 177:918–926. https://doi.org/10.1111/j.1469-8137.2007.02318.x

Colmer TD, Winkel A, Pedersen O (2011) A perspective on underwater photosynthesis in submerged terrestrial wetland plants. AoB Plants 2011. https://doi.org/10.1093/aobpla/plr030

Herzog M, Fukao T, Winkel A, Konnerup D, Lamichhane S, Alpuerto JB, Hasler-Sheetal H, Pedersen O (2018a) Physiology, gene expression and metabolome of two wheat cultivars with contrasting submergence tolerance. Plant Cell Environ 41:1632–1644. https://doi.org/10.1111/pce.13211

Herzog M, Konnerup D, Pedersen O, Winkel A, Colmer TD (2018b) Leaf gas films contribute to rice (*Oryza sativa*) submergence tolerance during saline floods. Plant Cell Environ 41:885–897. https://doi.org/10.1111/pce.12873

Huth MA, Huth A, Schreiber L, Koch K (2022) Design of a biomimetic, small-scale artificial leaf surface for the study of environmental interactions. Beilstein J Nanotechnol 13:944–957. https://doi.org/10.3762/bjnano.13.83

Koch K, Barthlott W (2009) Superhydrophobic and superhydrophilic plant surfaces: an inspiration for biomimetic materials. Philos Trans R Soc A-Math Phys Eng Sci 367:1487–1509. https://doi.org/10.1098/rsta.2009.0022

Konnerup D, Pedersen O (2017) Flood tolerance of *Glyceria fluitans* – the importance of cuticle hydrophobicity, permeability and leaf gas films for underwater gas exchange. Ann Bot 120:521–528. https://doi.org/10.1093/aob/mcx083

Konnerup D, Winkel A, Herzog M, Pedersen O (2017) Leaf gas film retentions during submergence of 14 cultivars of wheat (*Triticum aestivum*). Funct Plant Biol 44:877–887. https://doi.org/10.1071/FP16401

Kurokawa Y, Nagai K, Hung PD, Shimazaki K, Qu H, Mori Y, Toda Y, Kuroha T, Hayashi N, Aiga S, Itoh J-I, Yoshimora A, Sasaki-Sekimoto Y, Ohta H, Shimojima M, Malik AI, Pedersen O, Colmer TD, Ashikari M (2018) Rice leaf hydrophobicity and gas films are conferred by a wax synthesis gene (*LGF1*) and contribute to flood tolerance. New Phytol 218:1558–1569. https://doi.org/10.1111/nph.15070

Lauridsen T, Glavina K, Colmer TD, Winkel A, Irvine S, Lefmann K, Feidenhans'l R, Pedersen O (2014) Visualisation by high resolution synchrotron X-ray phase contrast micro-tomography of gas films on submerged superhydrophobic leaves. J Struct Biol 188:61–70. https://doi.org/10.1016/j.jsb.2014.08.003

Lu Y, Yu L, Zhang Z, Wu S, Li G, Wu P, Hu Y, Li J, Chu J, Wu D (2017) Biomimetic surfaces with anisotropic sliding wetting by energy-modulation femtosecond laser irradiation for enhanced water collection. RSC Adv 7:11170–11179. https://doi.org/10.1039/c6ra28174e

Mori Y, Kurokawa Y, Koike M, Malik AI, Colmer TD, Ashikari M, Pedersen O, Nagai K (2019) Diel O_2 dynamics in partially and completely submerged deepwater rice: leaf gas films enhance internodal O_2 status, influence gene expression, and accelerate stem elongation for 'snorkelling' during submergence. Plant Cell Physiol 65:973–985. https://doi.org/10.1093/pcp/pcz009

Neinhuis C, Barthlott W (1997) Characterization and distribution of water-repellent, self-cleaning plant surfaces. Ann Bot 79:667–677. https://doi.org/10.1006/anbo.1997.0400

Pedersen O, Colmer TD (2012) Physical gills prevent drowning of many wetland insects, spiders and plants. J Exp Biol 215:705–709. https://doi.org/10.1242/jeb.065128

Pedersen O, Rich SM, Colmer TD (2009) Surviving floods: leaf gas films improve O_2 and CO_2 exchange, root aeration, and growth of completely submerged rice. Plant J 58:147–156. https://doi.org/10.1111/j.1365-313X.2008.03769.x

Raskin I, Kende H (1983) How does deep water rice solve its aeration problem? Plant Physiol 72:447–454. https://doi.org/10.1104/pp.72.2.447

Raven JA (2008) Not drowning but photosynthesizing: probing plant plastrons. New Phytol 177:841–845

Spaeth M, Barthlott W (2008) Lotus-effect®: biomimetic super-hydrophobic surfaces and their application. Advances in Science and Technology

Teakle NL, Bowman S, Barrett-Lennard EG, Real D, Colmer TD (2012) Comparisons of annual pasture legumes in growth, ion regulation and root porosity demonstrate that *Melilotus siculus* has exceptional tolerance to combinations of salinity and waterlogging. Environ Exp Bot 77:175–184. https://doi.org/10.1016/j.envexpbot.2011.11.020

Teakle NL, Colmer TD, Pedersen O (2014) Leaf gas films delay salt entry and enhance underwater photosynthesis and internal aeration of *Melilotus siculus* submerged in saline water. Plant Cell Environ 37:2339–2349. https://doi.org/10.1111/pce.12269

Verboven P, Pedersen O, Ho QT, Nicolaï BM, Colmer TD (2014) The mechanism of improved aeration due to gas films on leaves of submerged rice. Plant Cell Environ 37:2433–2452. https://doi.org/10.1111/pce.12300

Winkel A, Colmer TD, Pedersen O (2011) Leaf gas films of *Spartina anglica* enhance rhizome and root oxygen during tidal submergence. Plant Cell Environ 34:2083–2092. https://doi.org/10.1111/j.1365-3040.2011.02405.x

Winkel A, Colmer TD, Ismail AM, Pedersen O (2013) Internal aeration of paddy field rice (*Oryza sativa* L.) during complete submergence – importance of light and floodwater O_2. New Phytol 197:1193–1203. https://doi.org/10.1111/nph.12048

Winkel A, Pedersen O, Ella ES, Ismail AM, Colmer TD (2014) Gas film retention and underwater photosynthesis during field submergence of four contrasting rice genotypes. J Exp Bot 65:3225–3233. https://doi.org/10.1093/jxb/eru166

Winkel A, Visser EJW, Colmer TD, Brodersen KP, Voesenek LACJ, Sand-Jensen K, Pedersen O (2016) Leaf gas films, underwater photosynthesis and plant species distributions in a flood gradient. Plant Cell Environ 39:1537–1548. https://doi.org/10.1111/pce.12717

Winkel A, Herzog M, Konnerup D, Floytrup AH, Pedersen O (2017) Flood tolerance of wheat – the importance of leaf gas films during complete submergence. Funct Plant Biol 44:888–898. https://doi.org/10.1071/FP16395

Wittig P, Ambros S, Müller JT, Bammer B, Konnerup D, Pedersen O, Mustroph A (2021) Two *Brassica napus* cultivars differ in gene expression, but not in their response to submergence. Physiol Plant 171:400–415. https://doi.org/10.1111/ppl.13251

Wu D, Wang JN, Wu SZ, Chen QD, Zhao S, Zhang H, Sun HB, Jiang L (2011) Three-level biomimetic rice-leaf surfaces with controllable anisotropic sliding. Adv Funct Mater 21:2927–2932

Yoo J-H, Park J-H, Cho S-H, Yoo S-C, Li J, Zhang H, Kim K-S, Koh H-J, Paek N-C (2011) The rice *bright green leaf* (*bgl*) locus encodes OsRopGEF10, which activates the development of small cuticular papillae on leaf surfaces. Plant Mol Biol 77:631–641

Zhang Z, Cheng Z-J, Gan L, Zhang H, Wu F-Q, Lin Q-B, Wang J-L, Wang J, Guo X-P, Zhang X, Zhao Z-C, Lei C-L, Zhu S-S, Wang C-M, Wan J-M (2016) *OsHSD1*, a hydroxysteroid dehydrogenase, is involved in cuticle formation and lipid homeostasis in rice. Plant Sci 249:35–45

Part IV
Aerenchyma Formation and Gas Transport

Chapter 7
Cavity Tissue for the Internal Aeration in Plants

Hirokazu Takahashi and Mikio Nakazono

7.1 Introduction

In wetland environments, soil pores are filled with water, and plant roots and microbes consume dissolved oxygen through respiration. Additionally, the rate of oxygen diffusion is 10,000 times lower in water than in air. Consequently, roots in waterlogged environments are deprived of oxygen (Jackson et al. 1985; Drew 1979). Effective internal aeration is essential for adaptation to wetland environments. Wetland plants have various adaptive characteristics that allow them to tolerate stresses associated with waterlogging (Bailey-Serres and Voesenek 2008; Voesenek and Bailey-Serres 2015). The important traits of tolerance to wetlands have been summarized by Colmer and Voesenek (2009); "aerenchyma" is one such important trait. In plants, aerenchyma is a tissue with internal air gaps for gaseous transport to sustain gas exchange (O_2, CO_2, ethylene, and methane) between shoots and roots (Armstrong 1979; Colmer 2003; Evans 2003; Jackson and Armstrong 1999). Aerenchyma has been observed in many plant species, and there are many types of aerenchyma with different morphological patterns (Fig. 7.1). These patterns were categorized into several groups based on the pattern of origin of aerenchyma and its systematic distribution (Justin and Armstrong 1987; Seago Jr et al. 2005). These types of aerenchyma is largely classified into primary and secondary types (Fig. 7.1; Jackson and Armstrong 1999; Takahashi et al. 2014). Primary aerenchyma is formed in primary tissues such as root and shoot cortical tissues. Secondary aerenchyma is newly differentiated tissue from the secondary meristem (i.e., phellogen or cork cambium) (Takahashi et al. 2014).

H. Takahashi · M. Nakazono (✉)
Graduate School of Bioagricultural Sciences, Nagoya University, Nagoya, Japan
e-mail: hiro_t@agr.nagoya-u.ac.jp; nakazono@agr.nagoya-u.ac.jp

J.-I. Sakagami, M. Nakazono (eds.), *Responses of Plants to Soil Flooding*,
https://doi.org/10.1007/978-981-99-9112-9_7

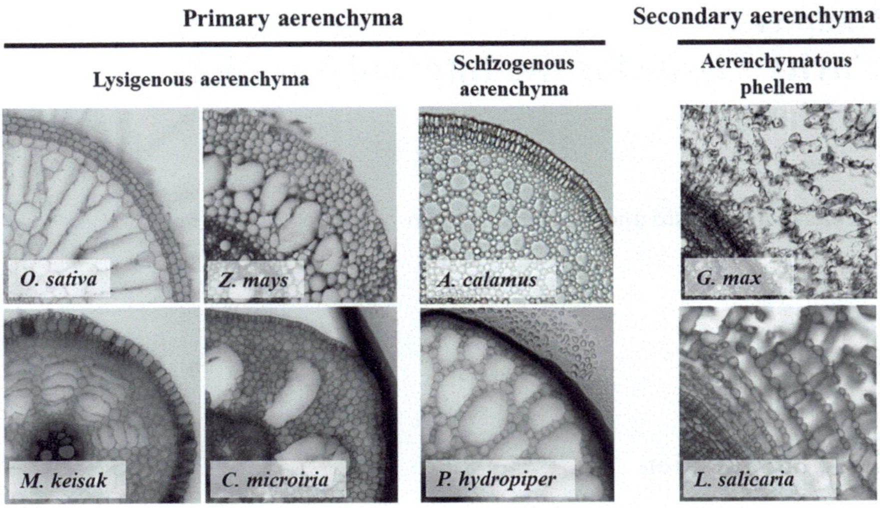

Fig. 7.1 Different patterns of aerenchyma in plants. The shape of the aerenchyma largely differs among different plant species. *Oryza sativa*, *Zea mays*, *Murdannia keisak*, and *Cyperus microiria* have lysigenous aerenchyma in their roots. *Acorus calamus* and *Persicaria hydropiper* have schizogenous aerenchyma in their roots. G*lycine max* and *Lythrum salicaria* form aerenchymatous phellem in the hypocotyls

7.2 Primary Aerenchyma

Seago Jr et al. (2005) classified root aerenchyma into seven groups (honeycomb expansigeny, radial expansigeny, schizogeny, schizo-lysigeny, packet lysigeny, tangential lysigeny, and radial lysigeny) based on the pattern of origin of aerenchymas and their systematic distribution. However, we discuss two categories, namely, schizogeny and lysigeny, in this chapter because separate evidence about each of the seven groups is still limited.

Schizogenous aerenchyma is a gas space between cells formed as a result of cell separation and expansion by differential cell division and/or cell expansion (Fig. 7.1; Arber 1920; Jackson and Armstrong 1999; Justin and Armstrong 1987; Takahashi et al. 2014). Schizogenous aerenchyma is constitutively formed not only in wetland plants but also in some non-wetland species (Evans 2003; Jackson and Armstrong 1999). The genus *Rumex* forms schizogenous aerenchyma in the taproots and adventitious roots (Colmer et al. 2004; Justin and Armstrong 1987; Laan et al. 1989). In *Rumex crispus* and *Rumex palustris*, the formation of schizogenous aerenchyma is promoted under waterlogged conditions (Colmer et al. 2004; Laan et al. 1989; Visser et al. 2000). *Sagittaria lancifolia* forms schizogenous aerenchyma in the petiole but lysigenous aerenchyma in the roots (Schussler and Longstreth 1996). Liang et al. (2008) showed that *S. trifolia* performs cell division at an early stage and cell enlargement in later stages to create the gas space, suggesting that cortical microtubule arrays play an important role in intercellular space formation, cell deformation, expansion, and deposition of secondary cell walls.

Lysigenous aerenchyma is a gas space formed as a result of death and subsequent lysis, causing the collapse of root cortical cells (Fig. 7.1; Arber 1920; Evans 2003; Jackson and Armstrong 1999; Justin and Armstrong 1987; Takahashi et al. 2014). Lysigenous aerenchyma is observed in the roots of wetland plants and many gramineous species. It is characterized by packet, tangential, and radial patterns. As shown in Fig. 7.1, *Cyperus microiria* forms packet lysigenous aerenchyma, and *Murdannia keisak* form tangential lysigenous aerenchyma. *Hydrocharis morsus-ranae* shows packet lysigeny patterns (Seago Jr et al. 1999, 2005), whereas *Cyperus alternifolius*, *Cladium jamaicense*, and *Carex nigra* show tangential lysigeny patterns (Justin and Armstrong 1987; Seago Jr et al. 2005). A radial lysigeny pattern is formed in rice (Jackson et al. 1985), maize (*Zea mays*; Drew et al. 1981), wheat (*Triticum aestivum*; Trought and Drew 1980), and barley (*Hordeum vulgare*; Arikado and Adachi 1955) and is well studied. During lysigenous aerenchyma formation in rice, cells in the mid-cortex die first, followed by the death of surrounding cortical cells of the roots (Kawai et al. 1998). In rice roots, cortical cells are completely digested, and only the remnants of collapsed cell walls are retained. The aerenchyma gradually enlarges from the apical to the basal part of the root. Aerenchyma is longitudinally interconnected along the root axis, and well-connected gas-filled pathways are visualized by X-ray computed tomography (Karahara et al. 2012).

7.3 "Constitutive" and "Inducible" Lysigenous Aerenchyma

Two types of lysigenous aerenchyma formation have been proposed: constitutive and inducible lysigenous aerenchyma. Both aerenchyma types are formed by cortical cell-specific programmed cell death (PCD). Constitutive lysigenous aerenchyma is aerobically formed along with root development, whereas inducible lysigenous aerenchyma is formed during waterlogging (Colmer and Voesenek 2009). Some wetland plants, such as rice, *Juncus effusus*, and *Zea nicaraguensis*, constitutively form lysigenous aerenchyma even under aerobic conditions and also form inducible lysigenous aerenchyma in response to waterlogging to enlarge gas spaces (Abiko et al. 2012; Jackson et al. 1985; Mano et al. 2006; Mano and Omori 2013; Visser and Bögemann 2006). In contrast, some non-wetland plant species such as maize, wheat, and barley do not form constitutive lysigenous aerenchyma but inducible lysigenous aerenchyma (McDonald et al. 2001; McPherson 1939; Trought and Drew 1980). The formation of inducible lysigenous aerenchyma requires several hours after waterlogging and low O_2 treatment even in wetland plants (Malik et al. 2003; Rajhi et al. 2011). Additionally, non-wetland plants have smaller aerenchyma than wetland plants (Armstrong 1979; Justin and Armstrong 1987; Colmer and Voesenek 2009). Rice forms constitutive lysigenous aerenchyma not only in the roots but also in the leaf sheaths, internodes, and leaf mid-ribs (Colmer and Pedersen 2008;

Matsukura et al. 2000; Parlanti et al. 2011; Steffens et al. 2011). The aerenchyma in shoots is connected to the root aerenchyma and enables gas exchange between shoots and roots, even in wetland environments. Thus, differences in the capacity of gas spaces may contribute to the waterlogging tolerance between wetland and non-wetland plants.

7.4 Molecular Mechanism of Lysigenous Aerenchyma Formation

The mechanisms of lysigenous aerenchyma formation have also been extensively studied. Recently, it has been reported that there are different mechanisms between constitutive and inducible lysigenous aerenchyma formation (Yamauchi and Nakazono 2022). The formation of both aerenchyma types is regulated by plant hormones; auxin and ethylene are involved in constitutive and inducible aerenchyma formation, respectively (Fig. 7.2). N-1-naphthyphthalamic acid (NPA) treatment suppresses aerenchyma formation under aerobic conditions, suggesting the inhibition of constitutive aerenchyma formation (Yamauchi et al. 2019). In line with this, the dominant negative mutant of the *auxin/indole-3-acetic acid (AUX/IAA) 13* (*iaa13*) gene shows reduced constitutive aerenchyma formation (Yamauchi et al. 2019). IAA13 directly binds to the auxin response factor (ARF) 19 and negatively regulates auxin signaling. Additionally, ARF19 directly binds to the auxin response element (AuxRE) in the promoter region of the gene encoding lateral organ boundary domain (LBD)-containing protein 1–8 and induces its expression. *LBD1–8* transcription is suppressed in *iaa13* mutant, but the restoration of *LBD1–8* expression recovers aerenchyma formation in *iaa13* mutant (Yamauchi et al. 2019). These results suggest that AUX/IAA-ARF-dependent auxin signaling regulates constitutive aerenchyma formation in rice (Fig. 7.2; Yamauchi et al. 2019). Constitutive aerenchyma is also formed in *Z. nicaraguensis*, which is a wild relative of maize (*Zea mays* ssp. *mays*) (Mano et al. 2007; Mano and Nakazono 2021), and is inhibited by NPA treatment (Ning et al. 2023). Interestingly, Ning et al. (2023) found that gravistimulation (root rotation) accelerated constitutive aerenchyma formation in the cortex on the concave side of the rotated roots. Gene expression analysis revealed that auxin levels were higher on the concave than on the convex side. These results suggest that AUX/IAA-ARF-dependent auxin signaling-dependent constitutive aerenchyma formation is conserved in several plants. Auxin is an important factor in constitutive aerenchyma formation, but it is not completely inhibited by NPA treatment or *iaa1*3 mutants. Mano et al. (2007) identified several genetic loci associated with constitutive aerenchyma in *Z. nicaraguensis*. This evidence suggests that constitutive aerenchyma formation is a quantitative trait and that regulatory mechanisms other than auxin levels control the process.

Ethylene induces the formation of lysigenous aerenchyma in maize, wheat, and rice (Drew et al. 1979, 2000; Justin and Armstrong 1991; Yamauchi et al. 2014).

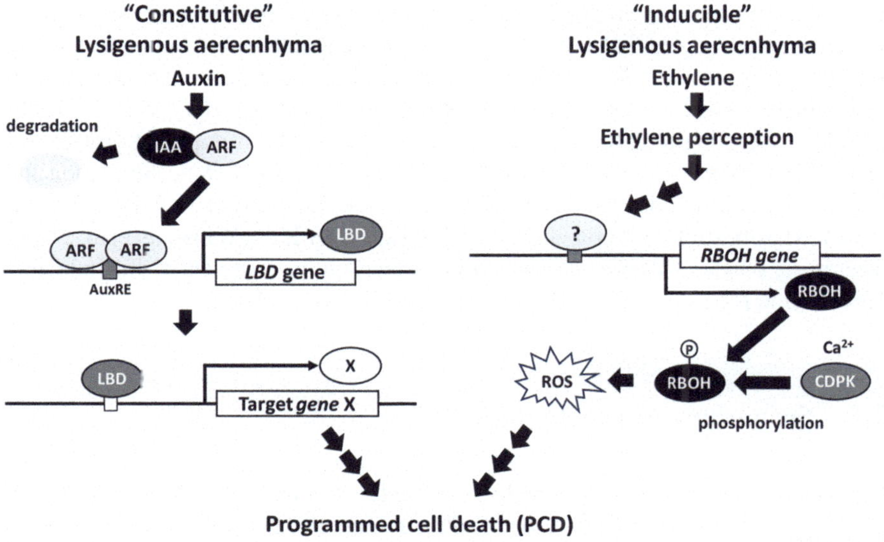

Fig. 7.2 Regulation of constitutive and inducible lysigenous aerenchyma formation by plant hormones. Auxins regulate constitutive aerenchyma. The IAA protein interacts with ARF, following which ARF cannot bind to the AuxRE in the promoter region of the *LBD* gene under auxin-absent conditions. Auxin promotes the degradation of IAA, and ARF can bind to AuxRE in the promoter region of *LBD*. LBD activates the expression of target genes and induces PCD. Ethylene regulates inducible aerenchyma formation. Ethylene production is promoted under low-oxygen conditions. After the perception of ethylene, expression of the *RBOH* gene is induced. RBOH is phosphorylated by CDPK for activation, and ROS is produced. ROS induces PCD in cortical cells

Ethylene is a gaseous plant hormone, and its accumulation increases in plant roots under waterlogged conditions (Visser and Voesenek 2004; Yamauchi et al. 2014; Yamauchi and Nakazono 2022). Ethylene biosynthesis genes are upregulated in response to waterlogging, and ethylene is a key factor in inducible aerenchyma formation.

Genes associated with lysigenous aerenchyma formation in maize roots have been identified using microarray analysis (Rajhi et al. 2011; Takahashi et al. 2015). Rajhi et al. (2011) found that a respiratory burst oxidase homolog (RBOH), a plant homolog of gp91^{phox} in mammalian NADPH oxidase, which plays a key role in reactive oxygen species (ROS) generation, was strongly upregulated in cortical cells (Rajhi et al. 2011; Yamauchi et al. 2011). Diphenyleneiodonium (DPI), an NADPH oxidase inhibitor, suppresses inducible aerenchyma formation in response to waterlogging and ethylene (Yamauchi et al. 2011; Takahashi et al. 2015). These results indicate that ROS are involved in aerenchyma formation downstream of ethylene signaling under waterlogged conditions (Fig. 7.2). This molecular mechanism of aerenchyma formation has been verified in rice roots, which form aerenchyma under well-drained aerobic conditions, and its formation enhances in response to waterlogged conditions. Rice also accumulates ethylene in the roots under waterlogged conditions, and the expression of the gene encoding RBOHH is induced by ethylene (Yamauchi et al. 2017). Knockout of the rice *RBOHH* gene significantly reduced waterlogging-induced H_2O_2 accumulation and aerenchyma formation. RBOH, whose activity is controlled by cytosolic Ca^{2+} influx from the apoplast and calcium-dependent protein kinase (CDPK), plays an important role in ROS-mediated signaling in defense responses, PCD, and plant development (Torres and Dangl 2005). The inhibition of cytosolic Ca^{2+} influx reduces H_2O_2 accumulation and aerenchyma formation (Yamauchi et al. 2017). Additionally, RBOHH activity is enhanced by phosphorylation via rice CDPK5 and CDPK13 (Yamauchi et al. 2017). Therefore, ethylene-ROS signaling during waterlogging is essential for inducible aerenchyma formation (Fig. 7.2; Yamauchi and Nakazono 2022).

7.5 Secondary Aerenchyma

Secondary aerenchyma is classified into the aerenchymatous phellem (AP), which is differentiated outward from the phellogen, and the porous secondary cortex, which is differentiated inward from the phellogen (Takahashi et al. 2014). Secondary aerenchyma is formed in the stems, adventitious roots, taproots, lateral roots, and nodules of several plants. Porous secondary cortices have been morphologically observed in several plants (Justin and Armstrong 1987; Gibberd et al. 2001), but molecular evidence is limited. AP has been well studied both morphologically and physiologically. AP is homologous to phellem (cork tissue), which is an impervious tissue that covers the plant surface. However, phellem usually consists of dead cells with suberized cell walls. In contrast, the AP consists of living cells with non-suberized cell walls and is thus distinguishable from cork tissues (Arber 1920; Fraser 1931; Jackson and Armstrong 1999). The AP is a highly porous tissue (Fig. 7.1) formed by successive cell divisions of phellogens, followed by the expansion of these cells (Arber 1920; Fraser 1931; Jackson and Armstrong 1999; Scott and Wager 1888). Therefore, the intercellular space of the AP steadily increases during development, and the volumes inside, which is in the vicinity of the phellogen, and outside, which is far from the phellogen, differ. The tissue porosity (i.e., the percentage of intercellular space) inside the AP was estimated to be approximately 50%, and that outside the AP was 80%; the volume of the intercellular space was greater outside than inside the AP (Takahashi et al. 2023). Morphologically, there are two types of AP; soybean (Fig. 7.1; Shimamura et al. 2003; Thomas et al. 2005), *Melilotus siculus* (Teakle et al. 2011), *Sesbania aculeata* (Scott and Wager 1888), and *Viminaria juncea* (Walker et al. 1983) have AP with radially elongated cells. In contrast, the AP of *Lythrum salicaria* is comprised of different cell types—one is alternating bands of radially elongated cells, and the other is isodiametric cells, which are adjacent to the phellogen and a band of non-elongated cells (Fig. 7.1).

AP contains a large volume of gas space, which enables internal aeration between the shoots and submerged roots. Interestingly, AP always forms immediately above the water level. Therefore, internal aeration is inhibited when the AP is completely submerged in water (Shimamura et al. 2010; Takahashi et al. 2023). This indicates that AP contributes to the entrance of oxygen uptake in addition to internal aeration. In legumes, AP facilitates not only oxygen but also nitrogen transport (Saraswati et al. 1992; Walker et al. 1983). Under waterlogged conditions, some legumes form AP on the surface of root nodules (Loureiro et al. 1994; Parsons and Day 1990; Shimamura et al. 2002; Thomas et al. 2005; Walker et al. 1983). Nodules, which are highly developed AP, can maintain nitrogen fixation even under waterlogged conditions (Bisseling et al. 1980; James et al. 1992).

According to Arber (1920), the term "aerenchyma" was defined in 1889, and "It was Schenck who proposed the useful term 'aerenchyma' for this non-suberised ventilating tissue produced by a phellogen." This suggests that "aerenchyma" first indicated secondary aerenchyma. Nonetheless, aerenchyma currently represents the primary aerenchyma (especially lysigenous aerenchyma), and the mechanism of secondary aerenchyma formation has not been well studied. In soybean, *L. salicaria*, and *V. juncea*, AP is not formed under well-drained soil conditions but is induced under waterlogged soil conditions (Shimamura et al. 2003; Stevens et al. 1997; Walker et al. 1983). Recently, there have been several reports of AP in soybeans. Soybean newly forms both phellogens and AP in response to waterlogged conditions (Shimamura et al. 2003; Takahashi et al. 2018). Takahashi et al. (2018) showed that AP development increased in a dose-dependent manner with day length, and phellogen and AP formation was completely inhibited under 24 h of darkness. Additionally, the formation of both tissues was inhibited under CO_2-deficit conditions. The sucrose concentration in hypocotyls was elevated under waterlogged and light conditions, but not under waterlogged and dark, drained and light, and drained and dark conditions. These results suggest that the allocation of photoassimilates, rather than local sucrose synthesis via starch degradation, is essential for phellogen and AP formation.

The regulatory mechanisms of phellogen and AP formation are still unknown; however, there have been few studies. Phellogen and AP formation was inhibited when the shoots above the epicotyl were excised, even under waterlogged conditions, but were recovered by the application of sucrose to the cut surface (Takahashi et al. 2018). Interestingly, this recovery occurred only with sucrose application, and not with glucose or fructose application. It is unclear why sucrose alone restored phellogen and AP formation in shoot-excised seedlings above the epicotyl. Sucrose acts as a signaling molecule in plant development; however, further studies are needed to determine whether sucrose acts as a regulator in phellogen and AP formation. Ethylene and auxin regulate the formation of lysigenous aerenchyma. In AP formation, anti-auxins, except *p*-chlorophenoxy isobutyric acid (PCIB), do not show any effect, and auxin levels are not altered under waterlogged conditions (Shimamura et al. 2015). Therefore, it remains unclear whether auxins are involved in the formation of phellogen and AP. In contrast, the abscisic acid (ABA) concentration in the hypocotyls gradually decreased to 50% within 24 h under waterlogged conditions. The exogenous application of ABA inhibits AP development in a dose-dependent manner. Interestingly, ABA did not inhibit phellogen differentiation and development but suppress the enlargement of phellogen-derived cells under waterlogged soil conditions (Shimamura et al. 2014). These results suggest that ABA is a negative regulator of AP differentiation from phellogen and/or AP development. To date, there is no additional evidence of AP formation.

In addition to the mechanisms underlying AP formation, there are some interesting questions regarding AP. The root of non-wetland plants radially leaks oxygen from the surface of the root. Some wetland plants that form primary aerenchyma prevent this radial oxygen loss (ROL) to maintain the oxygen inside roots (Nishiuchi et al. 2012). The components of the barrier to ROL are still unclear, but it is assumed that suberin and lignin contribute to the barrier to ROL. This type of barrier is required in AP because hypocotyl surfaces split and lenticels are formed during the formation of phellogens and AP. AP is directly exposed to water from lenticels under waterlogged soil conditions, and water can rapidly enter the intercellular spaces of well-developed AP. However, AP is hydrophobic and retains porous tissue even when submerged in water; the mechanism underlying the retention of tissue hydrophobicity is unclear. Takahashi et al. (2023) performed phellogen- and AP-specific microarray analyses using laser microdissection and found that the triterpenoid biosynthesis pathway is highly upregulated in AP. Consistent with gene expression, lupeol and betulinic acid, which are triterpenoids, accumulate in the hypocotyl under waterlogged conditions but not under drained conditions (Takahashi et al. 2023). Triterpenoids are plant metabolites derived from 2,3-oxidosqualene, a 30-carbon compound consisting of six isoprene units. 2,3-oxidosqualene is catalyzed by several oxidosqualene cyclases (OSCs). In sterol biosynthesis, 2,3-oxidosqualene is catalyzed by cycloartenol synthase (CAS), which is not highly expressed in AP. In the triterpenoid biosynthesis pathway, 2,3-oxidosqualene is converted to α-amyrin, β-amyrin, and lupeol by α-amyrin synthase (αAS), β-amyrin synthase (βAS), and lupeol synthase (LUS), respectively. These compounds are catalyzed by cytochrome P450 monooxygenases (P450s), which produce a wide variety of oxygenated triterpenoids (Thimmappa et al. 2014; Seki et al. 2015). The roles of these triterpenoids and their glycosides have been extensively studied in animal cells, as certain molecules exhibit anticancer, antibacterial, antifungal, and anti-inflammatory effects (Seki et al. 2008). However, there are few reports on the functions of triterpenoids in plants. Several triterpenoids are involved in plant pathogen defense, cell division, and cell elongation (Kemen et al. 2014; Misra et al. 2017; Singh et al. 2017). *Lotus japonicus* is suggested to be involved in nodule formation (Delis et al. 2011). However, the physiological and biological activities of plant triterpenoids remain largely unknown. Lupeol synthase mutants have been studied to determine the functions of lupeol and betulinic acid in AP. In the soybean genome, two copies of genes encoding LUS (i.e., *GmLUS1* and *GmLUS2*) and both proteins showed catalytic activity for LUS. A single mutant of *GmLUS1* (*lus1*) cannot accumulate lupeol and betulinic acid in the AP (Takahashi et al. 2023). Lupeol is a significant constituent of the epicuticular wax in the stem and hypocotyl of *Ricinus communis* (Guhling et al. 2006). On the surface of AP cells, the crystal structure of epicuticular wax is observed in the wild type but not in the *lus1* mutant (Fig. 7.3, Takahashi et al. 2023). Additionally, the tissue porosity of the AP in the *lus1* mutant is significantly lower than that in the wild type. These results suggest that triterpenoids contribute to tissue hydrophobicity by acting as epicuticular wax

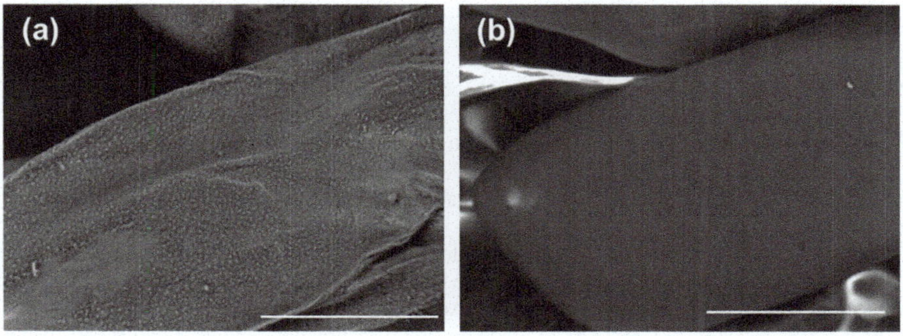

Fig. 7.3 Surface structure of AP cells. The cell surface of AP was observed using a low-vacuum pressure scanning electron microscope (LVP-SEM). Surface of AP cell in wild type (**a**) and *lus1* mutant (**b**). Bar = 20 μm

and preventing water penetration into the AP; this is because triterpenoids such as lupeol and betulinic acid are hydrophobic compounds. Loss of epicuticular wax and reduced porosity affect oxygen transport via the AP; consequently, the amount of oxygen delivered to the root tips is decreased in the *lus1* mutant. This reduction in oxygen transport results in shallow root systems under waterlogged conditions (Takahashi et al. 2023). These results indicate that triterpenoid accumulation in the AP contributes to effective internal aeration and root development for adaptation to waterlogging.

7.6 Perspective

Recently, there has been accumulating evidence on the molecular basis of aerenchyma formation, mainly lysigenous aerenchyma of Poaceae. Even in rice lysigenous aerenchyma, there are constitutive and inducible lysigenous aerenchyma with different regulatory mechanisms. Additionally, lysigenous aerenchyma is induced by not only waterlogging but also drought and nutrient deficiency (Lynch 2018; Jia et al. 2022; Schneider et al. 2023). It is unclear that the same mechanism in waterlogging contributes to the drought and nutrient deficiency-induced aerenchyma formation. Additionally, the mechanisms underlying secondary aerenchyma formation remain largely unknown. As shown in Fig. 7.1, the morphological patterns of aerenchyma are unique to plant species. At least, the formation of schizogenous and secondary aerenchyma should have different mechanisms of lysigenous aerenchyma formations because these aerenchyma were not formed by PCD. Further studies will be needed to understand all morphological patterns of aerenchyma formation in plants.

References

Abiko T, Kotula L, Shiono K, Malik AI, Colmer TD, Nakazono M (2012) Enhanced formation of aerenchyma and induction of a barrier to radial oxygen loss in adventitious roots of *Zea nicaraguensis* contribute to its waterlogging tolerance as compared with maize (*Zea mays* ssp. mays). Plant Cell Environ 35:1618–1630

Arber A (1920) Water plants: a study of aquatic angiosperms. Cambridge University Press, Cambridge

Arikado H, Adachi Y (1955) Anatomical and ecological responses of barley and some forage crops to the flooding treatment. Bull Fac Agric Mie Univ Tsu Mie 11:1–29

Armstrong W (1979) Aeration in higher plants. Adv Bot Res 7:236–332

Bailey-Serres J, Voesenek LACJ (2008) Flooding stress: acclimations and genetic diversity. Annu Rev Plant Biol 59:313–339

Bisseling T, Van Staveren W, Van Kammen A (1980) The effect of waterlogging on the synthesis of the nitrogenase components in bacteroids of *Rhizobium leguminosarum* in root nodules of *Pisum sativum*. Biochem Biophys Res Commun 93:687–693

Colmer TD (2003) Long-distance transport of gases in plants: a perspective on internal aeration and radial oxygen loss from roots. Plant Cell Environ 26:17–36

Colmer TD, Pedersen O (2008) Oxygen dynamics in submerged rice (*Oryza sativa*). New Phytol 178:326–334

Colmer TD, Peeters AJM, Wagemaker CAM, Vriezen WH, Ammerlaan A, Voesenek LACJ (2004) Expression of α-expansin genes during root acclimations to O_2 deficiency in *Rumex palustris*. Plant Mol Biol 56:423–437

Colmer TD, Voesenek LACJ (2009) Flooding tolerance: suites of plant traits in variable environments. Funct Plant Biol 36:665–681

Delis C, Krokida A, Georgiou S, Peña-Rodríguez LM, Kavroulakis N, Ioannou E, Roussis V, Osbourn AE, Papadopoulou KK (2011) Role of lupeol synthase in Lotus japonicus nodule formation. New Phytol 189:335–346

Drew MC, Jackson MB, Giffard S (1979) Ethylene-promoted adventitious rooting and development of cortical air spaces (aerenchyma) in roots may be adaptive responses to flooding in *Zea mays* L. Planta 147:83–88

Drew MC, Jackson MB, Giffard SC, Campbell R (1981) Inhibition by silver ions of gas space (aerenchyma) formation in adventitious roots of *Zea mays* L. subjected to exogenous ethylene or to oxygen deficiency. Planta 153:217–224

Drew MC, He CJ, Morgan PW (2000) Programmed cell death and aerenchyma formation in roots. Trends Plant Sci 5:123–127

Evans DE (2003) Aerenchyma formation. New Phytol 161:35–49

Fraser L (1931) The reaction of *Viminaria denudata* to increased water content of the soil. Proc Linn Soc N S W56:391–406

Gibberd MR, Gray JD, Cocks PS, Colmer TD (2001) Waterlogging tolerance among a diverse range of *Trifolium* accessions is related to root porosity, lateral root formation and 'aerotropic rooting'. Ann Bot 88:579–589

Guhling O, Hobl B, Yeats T, Jetter R (2006) Cloning and characterization of a lupeol synthase involved in the synthesis of epicuticular wax crystals on stem and hypocotyl surfaces of Ricinus communis. Archi Biochem Biophysic 448:60–72

Jackson MB, Armstrong W (1999) Formation of aerenchyma and the processes of plant ventilation in relation to soil flooding and submergence. Plant Biol 1:274–287

Jackson MB, Fenning TM, Jenkins W (1985) Aerenchyma (gas-space) formation in adventitious roots of rice (*Oryza sativa* L.) is not controlled by ethylene or small partial pressures of oxygen. J Exp Bot 36:1566–1572

James EK, Minchin FR, Sprent JI (1992) The physiology and nitrogen-fixing capability of aquatically and terrestrially grown *Neptunia plena*: the importance of nodule oxygen supply. Ann Bot 69:181–187

Jia X, Wu G, Strock C, Li L, Dong S, Zhang J, Zhao B, Lynch JP, Liu P (2022) Correction to: root anatomical phenotypes related to growth under low nitrogen availability in maize (Zea mays L.) hybrids. Plant Soil 477:843–844

Justin SHFW, Armstrong W (1987) The anatomical characteristics of roots and plant response to soil flooding. New Phytol 106:465–495

Justin SHFW, Armstrong W (1991) Evidence for the involvement of ethane in aerenchyma formation in adventitious roots of rice (*Oryza sativa* L.). New Phytol 118:49–62

Karahara I, Umemura K, Soga Y, Akai Y, Bando T, Ito Y, Tamaoki D, Uesugi K, Abe J, Yamauchi D, Mineyuki Y (2012) Demonstration of osmotically dependent promotion of aerenchyma formation at different levels in the primary roots of rice using a 'sandwich' method and X-ray computed tomography. Ann Bot 110:503–509

Kawai M, Samarajeewa PK, Barrero RA, Nishiguchi M, Uchimiya H (1998) Cellular dissection of the degradation pattern of cortical cell death during aerenchyma formation of rice roots. Planta 204:277–287

Kemen AC, Honkanen S, Melton RE, Findlay KC, Mugford ST, Hayashi K, Haralampidis K, Rosser SJ, Osbourn A (2014) Investigation of triterpene synthesis and regulation in oats reveals a role for beta-amyrin in determining root epidermal cell patterning. Proc Natl Acad Sci U S A 111:8679–8684

Laan P, Berrevoets MJ, Lythe S, Armstrong W, Blom CWPM (1989) Root morphology and aerenchyma formation as indicators of the flood-tolerance of *Rumex* species. J Ecol 77:693–703

Liang F, Shen LZ, Chen M, Yang Q (2008) Formation of intercellular gas space in the diaphragm during the development of aerenchyma in the leaf petiole of *Sagittaria trifolia*. Aquat Bot 88:185–195

Loureiro MF, De Faria SM, James EK, Pott A, Franco AA (1994) Nitrogen-fixing stem nodules of the legume, *Discolobium pulchellum* Benth. New Phytol 128:283–295

Lynch JP (2018) Rightsizing root phenotypes for drought resistance. J Exp Bot 69:3279–3292

Malik AI, Colmer TD, Lambers H, Schortemeyer M (2003) Aerenchyma formation and radial O_2 loss along adventitious roots of wheat with only the apical root portion exposed to O_2 deficiency. Plant Cell Environ 26:1713–1722

Mano Y, Omori F, Takamizo T, Kindiger B, Bird RM, Loaisiga CH, Takahashi H (2007) TL mapping of root aerenchyma formation in seedlings of a maize × rare teosinte "*Zea nicaraguensis*" cross. Plant Soil 295:103–113

Mano Y, Omori F (2013) Flooding tolerance in interspecific introgression lines containing chromosome segments from teosinte (*Zea nicaraguensis*) in maize (*Zea mays* subsp. mays). Ann Bot 112:1125–1139

Mano Y, Omori F, Takamizo T, Kindiger B, Bird R, Loaisiga CH (2006) Variation for root aerenchyma formation in flooded and non-flooded maize and teosinte seedlings. Plant Soil 281:269–279

Mano Y, Nakazono M (2021) Genetic regulation of root traits for soil flooding tolerance in genus *Zea*. Breed Sci 71:30–39

Matsukura C, Kawai M, Toyofuku K, Barrero RA, Uchimiya H, Yamaguchi J (2000) Transverse vein differentiation associated with gas space formation-fate of the middle cell layer in leaf sheath development of rice. Ann Bot 85:19–27

McDonald MP, Galwey NW, Colmer TD (2001) Waterlogging tolerance in the tribe Triticeae: the adventitious roots of *Critesion marinum* have a relatively high porosity and a barrier to radial oxygen loss. Plant Cell Environ 24:585–596

McPherson DC (1939) Cortical air spaces in the roots of *Zea mays* L. New Phytol 38:190–202

Misra RC, Sharma S, Garg A, Chanotiya CS, Ghosh S (2017) Two CYP716A subfamily cytochrome P450 monooxygenases of sweet basil play similar but nonredundant roles in ursane- and oleanane-type pentacyclic triterpene biosynthesis. New Phytol 214:706–720

Ning J, Yamauchi T, Takahashi H, Omori F, Mano Y, Nakazono M (2023) Asymmetric auxin distribution establishes a contrasting pattern of aerenchyma formation in the nodal roots of *Zea nicaraguensis* during gravistimulation. Front Plant Sci 14:1133009

Nishiuchi S, Yamauchi T, Takahashi H, Kotula L, Nakazono M (2012) Mechanisms for coping with submergence and waterlogging in rice. Rice 5:2

Parlanti S, Kudahettige NP, Lombardi L, Mensuali-Sodi A, Alpi A, Perata P, Pucciariello C (2011) Distinct mechanisms for aerenchyma formation in leaf sheaths of rice genotypes displaying a quiescence or escape strategy for flooding tolerance. Ann Bot 107:1335–1343

Parsons R, Day DA (1990) Mechanism of soybean nodule adaptation to different oxygen pressures. Plant Cell Environ 13:501–512

Rajhi I, Yamauchi T, Takahashi H, Nishiuchi S, Shiono K, Watanabe R, Mliki A, Nagamura Y, Tsutsumi N, Nishizawa NK, Nakazono M (2011) Identification of genes expressed in maize root cortical cells during lysigenous aerenchyma formation using laser microdissection and microarray analyses. New Phytol 190:351–368

Saraswati R, Matoh T, Sekiya J (1992) Nitrogen fixation of *Sesbania rostrata*: contribution of stem nodules to nitrogen acquisition. Soil Sci Plant Nutr 38:775–780

Schussler EE, Longstreth DJ (1996) Aerenchyma develops by cell lysis in roots and cell separation in leaf petioles in *Sagittaria lancifolia* (Alismataceae). Am J Bot 83:1266–1273

Schneider HM, Lor VS, Zhang X, Saengwilai P, Hanlon MT, Klein SP, Davis JL, Borkar AN, Depew CL, Bennett MJ, Kaeppler SM, Brown KM, Bhosale R, Lynch JP (2023) Transcription factor bHLH121 regulates root cortical aerenchyma formation in maize. Proc Natl Acad Sci U S A 120:e2219668120

Scott DH, Wager H (1888) On the floating-root of *Sesbania aculeata*. Pers Ann Bot 1:308–314

Seago JL Jr, Marsh LC, Stevens KJ, Soukup A, Votrubova O, Enstone DE (2005) A re-examination of the root cortex in wetland flowering plants with respect to aerenchyma. Ann Bot 96:565–579

Seago JL Jr, Peterson CA, Enstone DE (1999) Cortical ontogeny in roots of the aquatic plant, *Hydrocharis morsus-ranae* L. Can J Bot 77:113–121

Seki H, Ohyama K, Sawai S, Mizutani M, Ohnishi T, Sudo H, Akashi T, Aoki T, Saito K, Muranaka T (2008) Licorice beta-amyrin 11-oxidase, a cytochrome P450 with a key role in the biosynthesis of the triterpene sweetener glycyrrhizin. Proc Natl Acad Sci USA 105:14204–14209

Seki H, Tamura K, Muranaka T (2015) P450s and UGTs: key players in the structural diversity of triterpenoid saponins. Plant Cell Physiol 56:1463–1471

Shimamura S, Mochizuki T, Nada Y, Fukuyama M (2002) Secondary aerenchyma formation and its relation to nitrogen fixation in root nodules of soybean plants (*Glycine max*) grown under flooded conditions. Plant Prod Sci 5:294–300

Shimamura S, Mochizuki T, Nada Y, Fukuyama M (2003) Formation and function of secondary aerenchyma in hypocotyl, roots and nodules of soybean (*Glycine max*) under flooded conditions. Plant Soil 251:351–359

Shimamura S, Nishimura T, Koshiba T, Yamamoto R, Hiraga S, Nakamura T, Komatsu S (2015) Effects of anti-auxins on secondary aerenchyma formation in flooded soybean hypocotyls. Plant Prod Sci 19:154–160

Shimamura S, Yamamoto R, Nakamura T, Shimada S, Komatsu S (2010) Stem hypertrophic lenticels and secondary aerenchyma enable oxygen transport to roots of soybean in flooded soil. Ann Bot 106:277–284

Shimamura S, Yoshioka T, Yamamoto R, Hiraga S, Nakamura T, Shimada S, Komatsu S (2014) Role of abscisic acid in flood-induced secondary aerenchyma formation in soybean (*Glycine max*) hypocotyls. Plant Prod Sci 17:131–137

Singh AK, Kumar SR, Dwivedi V, Rai A, Pal S, Shasany AK, Nagegowda DA (2017) A WRKY transcription factor from Withania somnifera regulates triterpenoid withanolide accumulation and biotic stress tolerance through modulation of phytosterol and defense pathways. New Phytol 215:1115–1131

Steffens B, Geske T, Sauter M (2011) Aerenchyma formation in the rice stem and its promotion by H_2O_2. New Phytol 190:369–378

Stevens KJ, Peterson RL, Stephenson GR (1997) Morphological and anatomical responses of *Lythrum salicaria* L. (purple loosestrife) to an imposed water gradient. Int J Plant Sci 158:172–183

Takahashi H, Yamauchi T, Colmer TD, Nakazono M. (2014) Aerenchyma formation in plants. In: Plant cell monographs, Van Dongen JT Licausi F, ed. Springer: 247–265

Takahashi H, Yamauchi T, Rajhi I, Nishizawa NK, Nakazono M (2015) Transcript profiles in cortical cells of maize primary root during ethylene-induced lysigenous aerenchyma formation under aerobic conditions. Ann Bot 115:879–894

Takahashi H, Xiaohua Q, Shimamura S, Yanagawa A, Hiraga S, Nakazono M (2018) Sucrose supply from leaves is required for aerenchymatous phellem formation in hypocotyl of soybean under waterlogged conditions. Ann Bot 121:723–732

Takahashi H, Abo C, Suzuki H, Romusuk J, Oi T, Yanagawa A, Gorai T, Tomisaki Y, Jitsui M, Shimamura S, Mori H, Kaga A, Ishimoto M, Seki H, Miranaka T, Nakazono M (2023) Triterpenoids in Aerenchymatous Phellem contribute to internal root aeration and waterlogging adaptability in soybeans. New Phytol 239:936–948. https://doi.org/10.1111/nph.18975

Teakle NL, Armstrong J, Barrett-Lennard EG, Colmer TD (2011) Aerenchymatous phellem in hypocotyl and roots enables O_2 transport in *Melilotus siculus*. New Phytol 190:340–350

Thimmappa R, Geisler K, Louveau T, O'Maille P, Osbourn A (2014) Triterpene biosynthesis in plants. Ann Rev Plant Biol 65:225–257

Thomas AL, Guerreiro SMC, Sodek L (2005) Aerenchyma formation and recovery from hypoxia of the flooded root system of nodulated soybean. Ann Bot 96:1191–1198

Torres MA, Dangl JL (2005) Functions of the respiratory burst oxidase in biotic interactions, abiotic stress and development. Curr Opin Plant Biol 8:397–403

Trought MCT, Drew MC (1980) The development of waterlogging damage in young wheat plants in anaerobic solution cultures. J Exp Bot 31:1573–1585

Visser EJW, Bögemann GM (2006) Aerenchyma formation in the wetland plant *Juncus effusus* is independent of ethylene. New Phytol 171:305–314

Visser EJW, Colmer TD, Blom CWPM, Voesenek LACJ (2000) Changes in growth, porosity, and radial oxygen loss from adventitious roots of selected mono- and dicotyledonous wetland species with contrasting types of aerenchyma. Plant Cell Environ 23:1237–1245

Visser EJW, Voesenek LACJ (2004) Acclimation to soil flooding - sensing and signal-transduction. Plant Soil 254:197–214

Voesenek LACJ, Bailey-Serres J (2015) Flood adaptive traits and processes: an overview. New Phytol 206:57–73

Walker BA, Pate JS, Kuo J (1983) Nitrogen fixation by nodulated roots of *Viminaria juncea* (Schrad. & Wendl.) Hoffmans. (Fabaceae) when submerged in water. Aust J Plant Physiol 10:409–421

Yamauchi T, Rajhi I, Nakazono M (2011) Lysigenous aerenchyma formation in maize root is confined to cortical cells by regulation of genes related to generation and scavenging of reactive oxygen species. Plant Signal Behav 6:759–761

Yamauchi T, Watanabe K, Fukazawa A, Mori H, Abe F, Kawaguchi K, Oyanagi A, Nakazono M (2014) Ethylene and reactive oxygen species are involved in root aerenchyma formation and adaptation of wheat seedlings to oxygen-deficient conditions. J Exp Bot 65:261–273

Yamauchi T, Yoshioka M, Fukazawa A, Mori H, Nishizawa NK, Tsutsumi N, Yoshioka H, Nakazono M (2017) An NADPH oxidase RBOH functions in rice roots during lysigenous aerenchyma formation under oxygen-deficient conditions. Plant Cell 29:775–790

Yamauchi T, Tanaka A, Inahashi H, Nishizawa NK, Tsutsumi N, Inukai Y, Nakazono M (2019) Fine control of aerenchyma and lateral root development through AUX/IAA- and ARF-dependent auxin signaling. Proc Natl Acad Sci U S A 116:20770–20775

Yamauchi T, Nakazono M (2022) Mechanisms of lysigenous aerenchyma formation under abiotic stress. Trends Plant Sci 27:13–15

Chapter 8
Development and Regulation of a Radial Oxygen Loss Barrier to Acclimate to Anaerobic Conditions

Katsuhiro Shiono and Mikio Nakazono

8.1 Introduction

Plants can suffer from hypoxia or even anoxia when soils become waterlogged (i.e., only the root system is immersed in water). Well-drained soil is porous and normally filled with gas; excess water fills the pores, preventing atmospheric oxygen entry, as oxygen's diffusivity in water is approximately 10,000 times slower than in the air (Jackson et al. 1985). Other problems associated with waterlogging are the accumulation of phytotoxic compounds in the soil (Ponnamperuma 1984; Ernst 1990; Lamers et al. 1998; Kreuzwieser et al. 2004) and a decline in the availability of some nutrients (e.g., NO_3^- and Fe^{3+}) (Laanbroek 1990). Waterlogging damages many crops, including soybean, wheat, maize, oats, and barley (Setter and Waters 2003; Bertholdsson 2013; Shiono et al. 2019). Moreover, the number of flooding events is increasing worldwide due to climate change (Hirabayashi et al. 2013, 2021), causing severe damage to crop yields (Pedersen et al. 2017; Fukao et al. 2019).

Plants can adapt to waterlogging-induced hypoxia by shifting metabolic pathways (Bailey-Serres and Voesenek 2008; Fukao et al. 2019) and reducing energy consumption (Bailey-Serres and Voesenek 2010; Fukao et al. 2019). When waterlogging is prolonged, some plants can survive by internal oxygen transport via aerenchyma, which is an interior gas space connected between the shoot and the root tip (Colmer 2003b; Seago et al. 2005; Shiono et al. 2008; Colmer and Voesenek 2009; Yamauchi et al. 2018; Pedersen et al. 2021). A barrier to radial oxygen loss (ROL) and aerenchyma contributes to long-distance oxygen transport and

K. Shiono (✉)
Graduate School of Bioscience and Biotechnology, Fukui Prefectural University, Fukui, Japan
e-mail: shionok@fpu.ac.jp

M. Nakazono
Graduate School of Bioagricultural Sciences, Nagoya University, Nagoya, Japan

J.-I. Sakagami, M. Nakazono (eds.), *Responses of Plants to Soil Flooding*,
https://doi.org/10.1007/978-981-99-9112-9_8

waterlogging tolerance. Here we summarize how ROL barriers affect root growth and how their formation is regulated.

8.2 The Physiological Function of a Barrier to Radial Oxygen Loss (ROL)

8.2.1 Long-Distance Oxygen Transport for Respiration at the Root Tip

Under waterlogged conditions, oxygen molecules diffusing longitudinally through aerenchyma toward the root apex may be either consumed by respiration of root cells or diffused radially to the rhizosphere (Armstrong 1979; Colmer 2003b; Nishiuchi et al. 2012; Ejiri et al. 2021). Oxygen moving toward the root apex via the aerenchyma can leak into the surrounding soil (Armstrong 1979; Colmer and Voesenek 2009) in a process called radial oxygen loss (ROL). The amount of ROL is determined by the oxygen concentration gradient, the physical resistance to radial oxygen diffusion between aerenchyma and soil, and the consumption of oxygen by cells along the radial diffusion path (Fig. 8.1a) (Armstrong 1979; Armstrong and Beckett 1987; Colmer 2003b; Nishiuchi et al. 2012; Ejiri et al. 2021). The size of the path transporting oxygen in the roots (i.e., aerenchyma-to-cortex ratio and root thickness) also affects the amounts of ROL (Yamauchi et al. 2019; Pedersen et al. 2021). ROL aerates the rhizosphere (Armstrong 1979), where it nourishes the oxidative microbial community (Martin et al. 2019) and detoxifies toxic reduced substances in the waterlogged soil (Colmer 2003b; Neubauer et al. 2007; Blossfeld et al. 2011; Nishiuchi et al. 2012; Martin et al. 2019). However, ROL reduces the oxygen supply to the root apex, preventing the roots from growing deeper into anaerobic soil (Armstrong 1979; Sorrell et al. 2000; Pedersen et al. 2021).

Some wetland species form an ROL barrier in the outer cell layers exterior to the aerenchyma (Fig. 8.1b) (Nishiuchi et al. 2012; Yamauchi et al. 2018). Radial oxygen profiles across root tissues obtained with an oxygen microelectrode show that the ROL barrier is located in the outer cell layers, including sclerenchyma, exodermis/hypodermis, and epidermis (Armstrong et al. 2000; De Simone et al. 2002; Soukup et al. 2007). An ROL barrier along the basal part of roots enhances longitudinal oxygen diffusion to the root apex. This enables cells at the root tips that are surrounded by anaerobic soil to continue respiring (Fig. 8.1b(i)) (Armstrong 1979; Colmer 2003b; Nishiuchi et al. 2012; Pedersen et al. 2021). Multiple mathematical models support the idea that an ROL barrier leads to increased root length (Armstrong 1979; Sorrell et al. 2000; Pedersen et al. 2021). For instance, the maximum root length in roots with an ROL barrier was about twice that of roots without an ROL barrier when the cortex-to-stele area ratio was set at 10 (Pedersen et al. 2021).

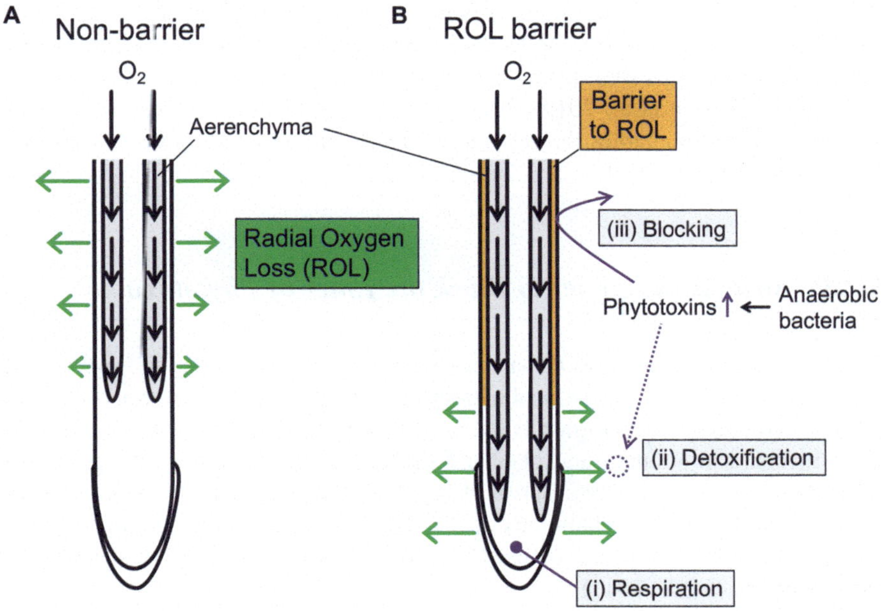

Fig. 8.1 Adaptive significance of an ROL barrier for roots in waterlogged soil. (**a**) Oxygen molecules diffusing longitudinally through aerenchyma toward the root apex may either be consumed by the respiration of root cells or diffuse radially to the rhizosphere (called radial oxygen loss, ROL). Although ROL aerates the rhizosphere, it also reduces the oxygen supply to the root apex. (**b**) Many wetland species form an ROL barrier in the basal parts of roots. ROL barrier promotes longitudinal oxygen diffusion by preventing losses to anaerobic soils. (i) In anaerobic soils, enhanced movement of oxygen toward the apex enables active respiration of the cells at the root tips. Because the root tip does not have an ROL barrier, (ii) the ROL around the root tip detoxifies toxic reduced substances in the waterlogged soil. (iii) The barrier also blocks the entry of potentially toxic compounds in highly reduced soils. [This figure from Ejiri et al. (2021) is reprinted with permission of *Breeding Science*]

8.2.2 Detoxification of Reduced Substances Around the Root Tip

In waterlogged soil, soil phytotoxins are produced by anaerobic bacteria (Fig. 8.1b). Because the root tip does not have an ROL barrier, the ROL around the root tip can neutralize (oxidize) toxic reduced substances such as sulfate and Fe^{2+} (Fig. 8.1b(ii)) (Armstrong 1979; Colmer 2003b; Yamauchi et al. 2018). Recent studies using a planar oxygen optode, an instrument that measures 2D in situ oxygen levels in vertical planes in the soil, solution, and rhizosphere (Jiménez et al. 2021b; Shiono et al. 2022a), have shown that light conditions dynamically change oxygen levels in the rhizosphere (Larsen et al. 2015; Lenzewski et al. 2018; Martin et al. 2019) and that Fe^{2+} oxidization was spatiotemporally associated with ROL (Maisch et al. 2019). The use of diffusive gradients in thin films (DGTs) is a technique for measuring 2D metal flux and concentrations at the surface of sediments (Beroyz et al. 1994;

Davison et al. 1997). A study using a planar oxygen optode combined with sulfide DGTs in the seagrasses *Halophila ovalis* and *Zostera muelleri* grown in marine sediments showed that ROL from the root tips reduces sulfate concentrations in the rhizosphere (Martin et al. 2019). This detoxification of reduced toxic substances around the root tip enables the roots to elongate into anaerobic soils (Armstrong 1979; Pedersen et al. 2021).

8.2.3 Structural Barrier to Block the Entry of Phytotoxins

Physical compounds of an ROL barrier have been debated. Lignin and suberin act as diffusion barriers in the extracellular spaces (apoplast) in the outer part of roots (OPRs) and endodermis. Lignin is a complex of polyphenolic polymers that provides mechanical strength and plant defense due to its resistance to degradation (Campbell and Sederoff 1996; Barros et al. 2015). Suberin is a hydrophobic macromolecule built from long-chain fatty acids, glycerol, and aromatic polymers (Graça 2015). Suberin lamellae are composed of suberin deposited on exodermal cell walls (Schreiber and Franke 2011).

When wetland plants, e.g., *Phragmites australis* (Soukup et al. 2007) and rice (Kotula et al. 2009), are grown under waterlogged or stagnant conditions for an extended period (i.e., 2–3 weeks), both suberized and lignified cell walls are frequently observed in the OPRs at the basal part of roots. Four lines of evidence show that suberin in the OPRs is a significant component of ROL barriers:

1. *Z. nicaraguensis* with ROL barrier develops suberin lamellae in the exodermis and lignin in the epidermis (Abiko et al. 2012; Watanabe et al. 2017). However, an introgression line of *Z. nicaraguensis* in a maize background (IL#468) with an ROL barrier under stagnant conditions develops a well-suberized exodermis/hypodermis along the basal parts of adventitious roots. Still, the epidermis is not lignified as it is in maize (inbred line Mi29) (Watanabe et al. 2017).
2. Roots of five wild *Echinochloa* species that constitutively form an ROL barrier under aerated conditions also have a well-suberized exodermis/hypodermis (Ejiri and Shiono 2019). However, two of the five species (a wetland ecotype *E. crus-galli* var. *crus-galli* and a non-wetland ecotype *E. colona*) do not have lignified sclerenchyma (Ejiri and Shiono 2019).
3. A transcriptome analysis using laser-microdissected tissues of the OPRs in rice showed that many genes involved in suberin biosynthesis (but not in lignin biosynthesis) are upregulated during ROL barrier formation (Shiono et al. 2014).
4. A metabolomic analysis in rice indicated that rice roots that form an ROL barrier accumulate malic acid and very-long-chain fatty acids, which are needed to synthesize suberin (Kulichikhin et al. 2014).

On the other hand, lignification in the OPRs is often unassociated with ROL barrier formation, depending on ecotypes, different developmental stages, or different genetic backgrounds (De Simone et al. 2003; Shiono et al. 2011; Watanabe et al.

2017; Ejiri and Shiono 2019; Ejiri et al. 2020). Therefore, the main component of an ROL barrier is thought to be suberin accumulated at the exodermis. Lignin alone seems not to form an ROL barrier. However, we should remark here the potential that lignin assists the function of suberization to impede oxygen diffusion along roots cannot be excluded.

The apoplast is a continuum of the extracellular space outside the plasma membrane (e.g., cell walls, xylem, and phloem), while the symplast is a continuum of intracellular space (e.g., cytosol and plasmodesmata). Roots take up water and ions via apoplastic and symplastic pathways. In general, suberin lamellae at the exodermis/hypodermis prevent water (Aloni et al. 1998; Kreszies et al. 2020), ions (Ranathunge et al. 2011a), and mycorrhizal fungi from moving through the apoplast (Enstone et al. 2003). In several kinds of grass that form an ROL barrier with well-suberized OPRs [e.g., rice (Ejiri et al. 2020), *P. australis* (Soukup et al. 2007), *Z. nicaraguensis* (Watanabe et al. 2017), wild *Echinochloa* species (Ejiri and Shiono 2019), and *O. glumaepatula* (W2165) (Ejiri et al. 2020)], tracers (e.g., periodic acids) were blocked from moving through the apoplast to the interior part of roots. Rice roots that form an ROL barrier can prevent Fe from entering the root via apoplast (Jiménez et al. 2021a), as well as the movements of H_2 and water vapor (Peralta Ogorek et al. 2021). Substances potentially toxic to plants (e.g., sulfate and Fe^{2+}) often accumulate in waterlogged soil (Ponnamperuma 1984; Ernst 1990; Lamers et al. 1998; Colmer 2003b; Kreuzwieser et al. 2004). Another benefit of ROL barriers in waterlogged soils is that their main component (suberin) can reduce the entry of such phytotoxins (Fig. 8.1b(iii)) (Ranathunge et al. 2011b; Watanabe et al. 2013; Kreszies et al. 2018; Yamauchi et al. 2018; Jiménez et al. 2021a; Peralta Ogorek et al. 2021).

8.2.4 Another Potential Component of an ROL Barrier

Regions that histochemically stain for suberin do not always function as ROL barriers. In maize, well-developed suberin lamellae were histochemically detected at the exodermis under stagnant conditions, but the roots were leaky to oxygen (Watanabe et al. 2017). Suberin lamellae in the OPRs were not histochemically visualized in the roots of *Halophila ovalis*, in which an ROL barrier was observed (Martin et al. 2019). In rice, ROL barrier formation was induced by 48 h of stagnant conditions, but suberin lamellae were not histochemically detected in the exodermis (Shiono et al. 2011). This failure to detect suberin may be due to a difference in suberin monomer composition and/or structure (Watanabe et al. 2017).

Rice appears to have another unidentified component that impedes oxygen diffusion. It is an electron-dense material that is packed in the intercellular space between the exodermis and the epidermis. It is observed in rice roots after 48 h of stagnant treatment using transmission electron microscopy (Shiono et al. 2011). These roots do not have histochemically stained suberin lamellae. Similar electron-dense materials are found in the intercellular spaces in the outer part of soybean nodules

(Parsons and Day 1990; James et al. 1991). The electron-dense material contains glycoproteins and forms an oxygen diffusion barrier (Parsons and Day 1990; James et al. 1991). This is necessary because the interior of the nodules must be anaerobic because nitrogenase activity that catalyzes the conversion of N_2 to ammonia is inactivated by oxygen (Denison et al. 1992; Denison 1998). Further studies are needed to identify the electron-dense components that inhibit oxygen transport in the OPRs.

8.3 Spatiotemporal Dynamics of Rhizosphere Oxidization

8.3.1 Effect of Root Length on ROL Barrier Formation

ROL barrier formation depends on the degree of soil waterlogging and the root developmental stage. In long roots of rice (ca. 110 mm), ROL barrier formation begins within 9 h of stagnant treatment and is completed within 24 h, while in short roots (ca. 60 mm), it takes 72–120 h to form a barrier (Shiono et al. 2011). Tissue age appears to affect ROL barrier formation because basal tissues of long roots are older than those of short roots, and suberin accumulation can be influenced by the developmental stage (Schreiber et al. 1999; Kotula et al. 2009; Serra and Geldner 2022). Root length also affects formation of the ROL barrier. Colmer et al. (2006) observed that an ROL barrier was formed along both long (>110 mm long) and short (ca. 60 mm) adventitious roots of rice grown in stagnant conditions for 3–4 weeks. In addition, only less than half of very short (<40 mm-long) adventitious roots had an ROL barrier (Ejiri et al. 2021). In the very short roots, an ROL barrier might not be essential for supplying sufficient oxygen for root tip respiration because the distance for oxygen diffusion from the root-shoot junction to the tips of the very short roots is much shorter than it is for long loots. After 5 days of incubation at 26 °C without plants, the agar solution at a depth of 5 mm was anoxic (0 mg/L) (Shiono, unpublished). We speculate that the emergence of roots into the anoxic and reduced solution induces barrier formation even in young adventitious roots.

In rice and common reed (*P. australis*), short lateral roots along adventitious roots leak oxygen due to the lack of an ROL barrier (Armstrong 1970; Armstrong et al. 1996; Colmer 2003b). Thus, the soil around the lateral roots and the very short adventitious roots was thought to be oxidized. Oxygen levels imaged with a planar oxygen optode indicated that the soil around the plant basal region, where many short lateral roots and very short roots are located, had high oxygen levels even though waterlogged soil without plants was anoxic (Larsen et al. 2015; Maisch et al. 2019). Figure 8.2 illustrates how an ROL barrier mediates oxygen distribution in the root system in soil-grown rice. Rice plants have many very short (<40 mm long) roots around the root-shoot junction, where new roots emerge (Fig. 8.2a). Although most short roots (between 40 and 80 mm) have an ROL barrier, very short roots (<40 mm) tend to lack an ROL barrier (Fig. 8.2b), thereby leading to oxygen

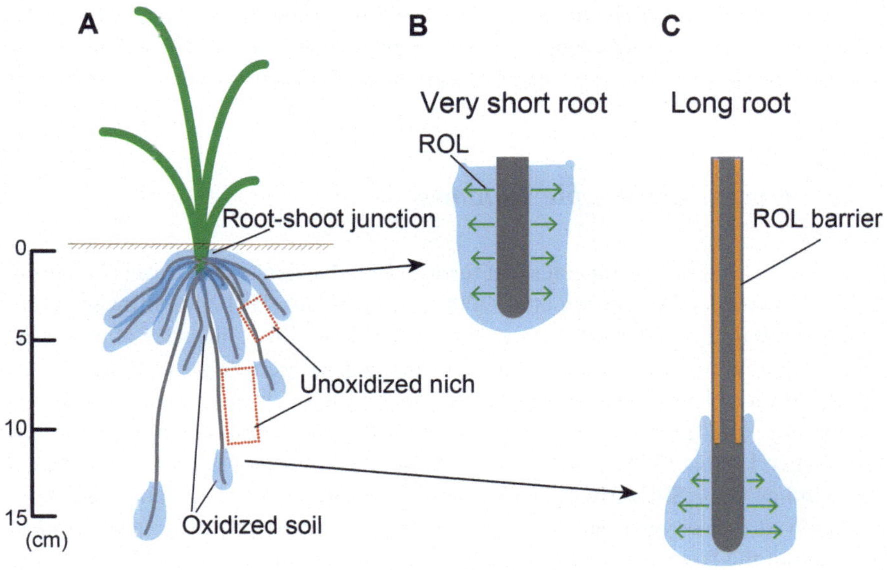

Fig. 8.2 Predicted locations of oxidized rhizosphere and ROL barrier formation along short and long roots. (**a**) Predicted locations of oxidized soil (blue) in the rhizosphere. (**b**, **c**) Radial oxygen loss (green arrows) in very short (<50 mm) roots (**b**) and long roots (>100 mm) (**c**). [This figure from Ejiri et al. (2021) is reprinted with permission of *Breeding Science*]

leakage and detoxification of reduced toxic substances to promote new root growth. On the other hand, most long roots form a strong ROL barrier along the subapical-to-basal regions, resulting in oxygen leakage only around the root tip (Fig. 8.2a, c). This allows long roots to serve as pioneers that detoxify reduced substances around the root tips, contributing to their active growth under waterlogging.

8.3.2 ROL Barrier Formation in Long Lateral Roots

In rice and *P. australis*, S-type lateral roots, which are short, thin, and incapable of branching, do not form an ROL barrier (Armstrong 1970; Armstrong et al. 1996; Colmer 2003b). Some plants also form L-type lateral roots, which are long, thick, and capable of branching. Under stagnant conditions, It is reported a formation of ROL barrier along the L-type lateral roots in rice (Noorrohmah et al. 2020) and *Zea nicaraguensis*, a wild relative of maize that grows in floodplains (Pedersen et al. 2021). *Z. nicaraguensis* has a sparse root system: Abiko et al. (2012) reported that 25-day-old *Z. nicaraguensis* grown under waterlogged soil for 21 days had only 15 adventitious roots. We speculate that the rhizosphere of *Z. nicaraguensis* has large unoxidized niches (rectangles in Fig. 8.2a) due to its sparse root system and that L-type lateral roots elongate into these niches. Like the ROL barrier of adventitious roots (Fig. 8.1), the ROL barrier of L-type lateral roots may help the roots to grow into the reduced soils in the unoxidized niches. Monitoring the oxygen status in the niches during lateral root elongation with a planar oxygen optode would clarify whether long lateral roots help adapt to waterlogged conditions.

8.3.3 Oxygen-Permeable Windows

In roots of several wetland species that form an ROL barrier, such as rice (Armstrong 1971), *P. australis* (Armstrong et al. 2000), *H. marinum* (H819) (Garthwaite et al. 2008), and wild *Echinochloa* species (Ejiri and Shiono 2019), the sites of the emergence of lateral roots contain passage cells that leak oxygen. Areas of passage cells where oxygen leaks and lateral roots emerge have been called "windows" (Armstrong et al. 2000). Armstrong et al. (2000), using a microelectrode, showed that oxygen levels in *P. australis* were higher at the surface of window sites than at non-window sites. Similarly, Ejiri and Shiono (2019), using methylene-blue staining, showed that oxygen was leaking in a spotty manner (probably from window sites) along the basal part of roots in *Echinochloa* accessions. Interestingly, in rice (Justin and Armstrong 1987), *P. australis* (Armstrong et al. 2000), and wild *Echinochloa* species (Ejiri and Shiono 2019, 2020), the exodermis at window sites lacks suberin lamellae, and the cortical cells are left intact without programmed cell death (i.e., without aerenchyma) (Fig. 8.3). However, the lateral root primordium does not reach the cortex and the exodermis. At window sites of *E. crus-galli*, lignin deposits at the sclerenchyma and the exodermis were also clearly reduced (Ejiri and Shiono 2020).

Passage cells possess Casparian strips composed of lignin and small deposits of the suberin (Schreiber and Franke 2011) but not suberin lamellae such as those covering the plasma membrane of the exodermal cells (Peterson and Enstone 1996). The development of exodermal passage cells is well described in maize (Enstone and Peterson 1997, 2005), although the oxygen permeability of the passage cells has not been evaluated. In the younger part of maize root, the exodermis matured in a patchy manner with some unsuberized cells, some developing suberin lamellae, and some mature suberin lamellae (Enstone and Peterson 2005). Exodermal cells surrounding emerged lateral roots remained unsuberized longer than other cells in the layer, thus becoming exodermal passage cells (Enstone and Peterson 2005). Applying toxic sulfide to rice roots induced unusually high suberin and/or lignin accumulations in the OPRs, preventing lateral roots from penetrating the sclerenchyma and growing through the cortex of adventitious roots (Armstrong and Armstrong 2005). This suggests that well-suberized and/or well-lignified cells in the OPRs block the emergence of lateral roots.

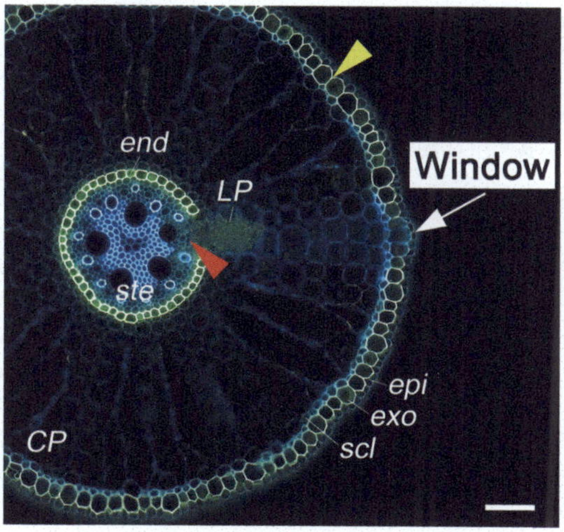

Fig. 8.3 An oxygen-permeable window site in the adventitious roots of *Echinochloa colona*. Cross-section of an adventitious root showing the location of a window (white arrow) at a site in the exodermis where a lateral root is expected to emerge. The window is composed of passage cells that lack suberin lamellae. Suberin lamellae are indicated as yellow-green fluorescence with Fluorol Yellow 088 (yellow arrowhead). Red arrowhead indicates the pericycle from which lateral root primordia are predicted to emerge. Blue fluorescence indicates autofluorescence. Fluorol Yellow 088 staining was conducted as described in Ejiri and Shiono (2019). The plants were grown in an aerated nutrient solution for 10 days and then transferred into a deoxygenated stagnant nutrient solution for 14 days. *CP* cortical parenchyma, *end* endodermis, *epi* epidermis, *exo* exodermis, *LP* lateral root primordia, *scl* sclerenchyma, *ste* stele. Scale bars: 100 μm. [This figure from Ejiri et al. (2021) is reprinted with permission of *Breeding Science*]

8.4 Distribution of Plant Species That Have ROL Barriers

8.4.1 Wetland and Non-wetland Plants

For the past two decades, many studies have evaluated ROL barrier formation in mono- and dicotyledonous species grown in wetland, non-wetland, and brackish water regions (Table 8.1). In general, ROL barriers do not form in waterlogging-sensitive non-wetland species including major crops, e.g., maize (*Z. mays* ssp. mays inbred maize line Mi29) (Abiko et al. 2012), wheat (*Triticum aestivum*) (McDonald et al. 2001b), barley (*Hordeum vulgare*) (McDonald et al. 2001a), *Brachypodium distachyon* (Shiono and Yamada 2014), *Brassica napus* (Voesenek et al. 1999), *Pisum sativum* (Healy and Armstrong 1972), and *Sorghum bicolor* (McDonald et al. 2002). Most waterlogging-tolerant species that can form an ROL barrier are distributed in wetlands (Table 8.1). Typical examples are rice and wild *Echinochloa* weeds that grow in paddy fields (Colmer et al. 1998; Ejiri and Shiono 2019), *P. australis* growing on riversides (Armstrong and Armstrong 2001), and *Z. nicaraguensis* and *Oryza glumaepatula* growing in floodplains (Abiko et al. 2012; Ejiri et al. 2020). However, ROL barriers were not found in some wetland species, including

Table 8.1 Plants species that have a tight barrier to ROL

Species	Habitat	Induction	References
Dicotyledonous			
Tabernaemontana juruana	W	Unknown	De Simone et al. (2002, 2003)
Caltha palustris	W	Inducible	Visser et al. (2000)
Bruguiera gymnorhiza	B	Unknown	Cheng et al. (2012)
Monocotyledonous			
Echinochloa colona	N	Constitutive	Ejiri and Shiono (2019)
E. crus-galli var. *praticola*	N	Constitutive	Ejiri and Shiono (2019)
Colocasia esculenta	N/W	Unknown	Abiko and Miyasaka (2020)
E. crus-galli var. *crus-galli*	N/W	Constitutive	Ejiri and Shiono (2019)
Cyperus eragrostis	W	Constitutive	Manzur et al. (2015)
Critesion marinum	W	Inducible	McDonald et al. (2001a)
E. crus-galli var. *formosensis*	W	Constitutive	Ejiri and Shiono (2019)
E. crus-galli var. *mitis*	W	Constitutive	McDonald et al. (2001a, 2002)
E. oryzicola	W	Inducible	Ejiri and Shiono (2019)
Glyceria maxima	W	Unknown	Soukup et al. (2007)
Hordeum marinum ssp. *marinum* (Acc. H21)	W	Inducible	Garthwaite et al. (2003) and Malik et al. (2011)
Juncus effusus	W	Constitutive	Visser et al. (2000)
Oryza glumaepatula	W	Constitutive	Ejiri et al. (2020)
O. rufipogon	W	Inducible	Ejiri et al. (2020)
O. sativa	W	Inducible	Colmer et al. (1998)
Phragmites australis	W	Unknown	Armstrong and Armstrong (2001)
Urochloa humidicola	W	Constitutive	Jiménez et al. (2019)
Zea nicaraguensis	W	Inducible	Abiko et al. (2012)
Halophila ovalis	B	Unknown	Connell et al. (1999)
Zostera muelleri	B	Unknown	Martin et al. (2019)

Induction of an ROL barrier: constitutive means an ROL barrier is formed under aerated (or well-drained) and stagnant (or waterlogged) conditions. Inducible means an ROL barrier is formed under stagnant (or waterlogged) conditions but is not formed under aerated (or well-drained) conditions. Unknown means an ROL barrier formed under stagnant (or waterlogged conditions) but that it was not evaluated under aerated (or well-drained) conditions. Thus, these species can form an ROL barrier, but it remains unclear whether it is constitutive. This table from Ejiri et al. (2021) is reprinted with permission of *Breeding Science*

W wetland, *N* non-wetland, *B* brackish water region, *N/W* non-wetland or wetland

Paspalidium geminatum (Manzur et al. 2015), *Ranunculus sceleratus* (Visser et al. 2000), *Rumex crispus* (Manzur et al. 2015), *R. palustris* (Visser et al. 2000), and *Salix martiana* (De Simone et al. 2002). This may be because these species live in areas where soil substrates are not highly reduced (e.g., low in organic matter; flowing water) and thus may not need an ROL barrier (Smits et al. 1990; Colmer 2003b). Alternatively, these species may possess genetic factor(s) related to tolerance to toxic soil constituents under flooded reducing soil conditions.

8.4.2 Constitutive and Inducible ROL Barrier

ROL barriers are classified as inducible and constitutive (Table 8.1). In some wetland species, including rice (Colmer et al. 1998), *H. marinum* (Garthwaite et al. 2003), *Z. nicaraguensis* (Abiko et al. 2012), and *Caltha palustris* (Visser et al. 2000), an ROL barrier is induced in roots by growth in waterlogged soil or stagnant conditions, while roots remain leaky to oxygen under well-drained soil or aerated conditions. ROL barriers form constitutively (i.e., in the absence of waterlogging) in several wetland species, including *Cyperus eragrostis* (Manzur et al. 2015), *Juncus effusus* (Visser et al. 2000), wild *Echinochloa* species (Ejiri and Shiono 2019), *O. glumaepatula* (Ejiri et al. 2020), and *Urochloa humidicola* (Jiménez et al. 2019) (Table 8.1). The ability to form a constitutive or inducible barrier is considered an adaptation to waterlogging (Colmer 2003b). The patterns of ROL barrier formation were evaluated in annual wild *Echinochloa* species (Ejiri and Shiono 2019). All three *Echinochloa crus-galli* varieties including var. *crus-galli*, *formosensis*, and *praticola* formed a constitutive ROL barrier, whereas *E. oryzicola*, like rice (Colmer 2003a), had an inducible ROL barrier (Ejiri and Shiono 2019). Phylogenic analyses suggest that *E. crus-galli* (which is hexaploid) is derived from *E. oryzicola* (which is tetraploid) (Aoki and Yamaguchi 2008, 2009). *E. oryzicola* is limited to waterlogged paddies, while *E. crus-galli* adapts to wet and dry areas depending on the variety (Yamasue 2001; Rao et al. 2007; Tanesaka et al. 2010). It is expected that constitutive suberization in the outer root cell layers that form an ROL barrier also increases drought and salinity tolerance (Enstone et al. 2003; Ranathunge et al. 2011b; Kreszies et al. 2018). The ability of *E. crus-galli* to grow in a wide range of habitats may be due to its constitutive barrier that increases its tolerance for multiple stresses.

8.5 Mechanisms of ROL Barrier Formation

8.5.1 Environmental Trigger

Stagnant or waterlogged conditions induce ROL barriers in some wetland species including rice (Colmer et al. 1998; Ejiri et al. 2021). How does waterlogging induce ROL barrier formation? Prolonged waterlogging causes the depletion of oxygen and the accumulation of other products, such as ethylene, CO_2, and phytotoxins (e.g., Fe^{2+}, NH_4^+ and organic acids) in the soil, as well as pH changes in the soil (i.e., a decrease in alkaline soil and an increase in acidic soil) (Ponnamperuma 1984). Low oxygen and high ethylene initiate aerenchyma formation, promoting longitudinal diffusion of oxygen (Shiono et al. 2008; Voesenek and Sasidharan 2013; Yamauchi et al. 2018). Interestingly, in rice, ROL barrier formation is not triggered by low oxygen, elevated ethylene and CO_2 (Colmer et al. 2006), or low pH (4.7 or 5.7) (Colmer et al. 2019). On the other hand, ROL barrier formation can be stimulated by soil phytotoxins including Fe^{2+} (Mongon et al. 2014), NH_4^+ (Ranathunge et al. 2016), sulfide (Armstrong and Armstrong 2005; Peralta Ogorek et al. 2023), and organic acids (Armstrong and Armstrong 2001; Kotula et al. 2014) that are produced by anaerobic bacteria in waterlogged soils. However, adding sulfide (0.174 mM) or organic acids (e.g., acetic acids 0.4–1.5 mM) to a hydroponic solution was found to inhibit root growth due to their toxicity and results in increased suberization and/or lignification of cell walls in the OPRs (Armstrong and Armstrong 2001, 2005; Kotula et al. 2014). Colmer et al. (2019) demonstrated that nontoxic levels (<0.05 mM) of organic acids such as acetic, propionic, butyric, and hexanoic acids initiate ROL barrier formation in rice roots, without affecting root tissue respiration. Mongon et al. (2014) also showed that nontoxic levels of Fe^{2+} (0.36–0.54 mM) induce an ROL barrier under aerated conditions, without inhibiting root growth. Thus, ROL barrier formation in rice is considered to be triggered by accumulations of soil phytotoxins in waterlogged soil (Pedersen et al. 2021).

8.5.2 Hormone Signaling and Molecular Mechanisms

Ethylene is a crucial hormone in acclimation to waterlogging. However, ethylene does not induce an ROL barrier in rice (Colmer et al. 2006). Plant hormones involved in an ROL barrier formation remain unknown for a long time. Recently, we identified abscisic acid (ABA) as an inducer of suberization in the exodermis to form an ROL barrier in rice (Shiono et al. 2022b). Inhibiting ABA biosynthesis by gene knockout (*osaba1*) or by adding ABA biosynthesis inhibitor suppressed the ROL barrier formation and exodermal suberization (Shiono et al. 2022b). Exogenous ABA strongly induced the formation of an ROL barrier in wild-type plants under aerated conditions and restored the barrier in osaba1 mutants with suberin lamellae in the exodermis (Shiono et al. 2022b). Endogenous ABA concentrations were slightly higher under stagnant conditions than under aerated conditions (Shiono et al. 2022b). Additionally, tissue-specific expression patterns of genes related to ABA biosynthesis and degradation suggest that the ABA level was upregulated at the ROL barrier formation (Shiono et al. 2022b). Thus, a model is proposed that ABA, in response to waterlogging, induces suberization in the root exodermis that results in an ROL barrier (Fig. 8.4) (Shiono et al. 2022b).

It is unclear how ABA mediates suberization in the exodermis, but there has been some progress in how it mediates suberization in the endodermis. In the root maturation zone of *Arabidopsis*, *SHORT ROOT* (*SHR*), a GRAS transcription factor, is known to promote endodermal differentiation, including Casparian strip formation (Drapek et al. 2018; Li et al. 2018). SHR also promotes suberin lamellae formation via enhancing ABA levels and *MYB36* (Wang et al. 2020). *MYB36* and the elevated ABA independently strengthen the expression of *MYB39* that directly promotes suberin biosynthesis genes (Wang et al. 2020). In the exodermis, 98 genes that were strongly upregulated during ROL barrier formation were identified by a transcriptome analysis using laser-microdissected tissues of the OPRs in rice (Shiono et al.

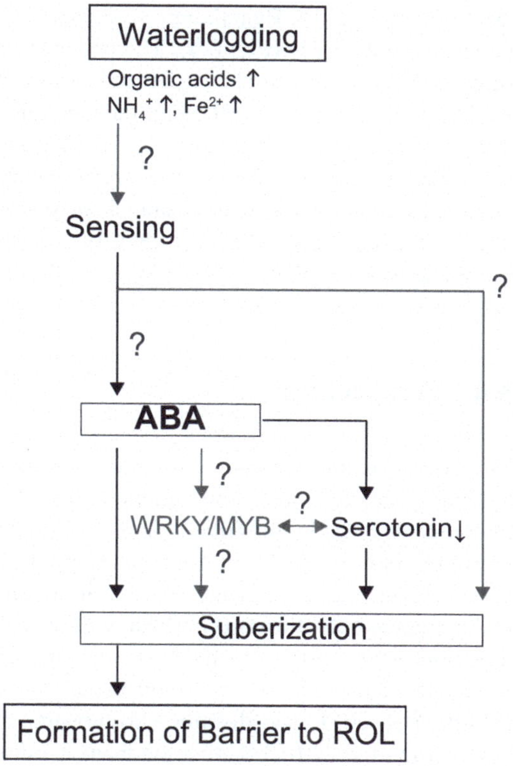

Fig. 8.4 Model of radial oxygen loss (ROL) barrier formation in rice roots under waterlogged conditions. Under stagnant conditions that mimic waterlogged soil, abscisic acid (ABA) worked as an inducer of suberin lamellae development in the exodermis, resulting in an ROL barrier formation. Arrows represent positive and negative regulation, respectively

2014). These included suberin biosynthesis genes such as *CYTOCHROME P450* (*OsCYP86B3*, *LOC_Os10g34480*) and *ABC TRANSPORTER* (*OsABCG5*, *LOC_Os03g17350*) and transcription factor genes containing WRKY, AP2, and MYB domains (Shiono et al. 2014). Additionally, many (65–78%) of the promoters of the upregulated transcription factors had putative *cis*-elements such as WRKY, AP2, NAC, and MYB (Shiono et al. 2014). The application of low concentrations of organic acids also upregulated two suberin biosynthesis genes (*OsABCG5* and *OsCYP86B3*) (Colmer et al. 2019). Lu et al. (2022) demonstrated that serotonin acts downstream of ABA signaling as a negative regulator of exodermal/endodermal suberization in rice under salt stress. In the exodermis, an increase in endodermal ABA or an altered sensitivity to ABA might act as signals that induce the expression of transcription factors (*WRKY* and *MYB*) and enhance suberin biosynthesis genes involvement of serotonin signaling (Fig. 8.4). Even in barley that does not form an ROL barrier (Garthwaite et al. 2003; McDonald et al. 2001a), barley can artificially induce an ROL barrier with suberized exodermis when ABA is exogenously applied (Shiono and Matsuura 2024). The resulting construction of suberin lamellae in the exodermis would decrease oxygen permeability.

ROL barrier formation in rice is considered to be triggered by the accumulation of soil phytotoxins that are produced by anaerobic bacteria in waterlogged soils

(Mongon et al. 2014; Ranathunge et al. 2016; Colmer et al. 2019; Ejiri et al. 2021; Pedersen et al. 2021) (Fig. 8.4). The endogenous ABA level is increased by exposure to NH_4^+ (Sun et al. 2020) and heavy metal (Bücker-Neto et al. 2017). Additionally, damage to cell membranes by the phytotoxins may induce the suberization of plant roots (Armstrong and Armstrong 2001; Watanabe et al. 2013). The signal that triggers an ABA or other signaling network to form an ROL barrier is not known. Further studies on the changes in gene expression in the roots immediately after waterlogging should help identify the trigger of ABA signaling that helps acclimate wetland plants to waterlogged conditions.

8.6 Perspectives

Most waterlogging-sensitive crops cannot form an ROL barrier and/or a significant amount of aerenchyma. Some attempts have been made to improve waterlogging tolerance in crops by introducing genes or quantitative trait loci (QTLs) associated with ROL barrier formation from waterlogging-tolerant wild relatives. *H. marinum*, a waterlogging-tolerant wild relative of wheat, inducibly forms an ROL barrier under stagnant conditions (Garthwaite et al. 2003). To obtain wheat varieties with inducible ROL barriers in the roots, wheat cultivars [e.g., Chinese Spring (CS)] were hybridized with *H. marinum* accessions (e.g., H21), producing amphiploid (Malik et al. 2011) and disomic chromosome addition lines for H21 in CS wheat (Konnerup et al. 2017). Under stagnant conditions, four of the amphiploid lines, including the H21-CS amphiploid, had tight ROL barriers (Malik et al. 2011). On the other hand, Konnerup et al. (2017) did not detect ROL barriers in any of the six disomic chromosome addition lines for chromosomes 1, 2, 4, 5, 6, or 7 of H21 in CS wheat. The gene involved in ROL barrier formation may be located on chromosome 3 of H21, but unfortunately, a disomic chromosome addition line of chromosome 3 of H21 was not isolated and could not be tested (Konnerup et al. 2017). The wild relative of maize, *Z. nicaraguensis*, generates an inducible ROL barrier under stagnant conditions, while the inbred maize line (Mi29) does not (Abiko et al. 2012). Analyses of *Z. nicaraguensis* introgression lines in the genetic background of domesticated maize identified a major locus associated with ROL barrier formation in a segment of the short arm of chromosome 3 of *Z. nicaraguensis* (Watanabe et al. 2017; Mano and Nakazono 2021). We also found that some accessions of a wild rice species, *O. glumaepatula*, have a constitutive ROL barrier with suberized exodermis under aerated conditions (Ejiri et al. 2020; Ejiri and Shiono 2023). *O. glumaepatula* has an AA genome, making it possible to cross it with cultivated rice that does not form a constitutive ROL barrier. We are presently mapping ROL-related QTLs in introgression lines of *O. glumaepatula* to identify the genes that regulate constitutive ROL barrier formation.

Acknowledgments We thank Mr. Masato Ejiri for the discussion and excellent comments on the draft of this manuscript. This work was partly supported by JSPS KAKENHI (JP16KK0173, JP19K05978 and JP22K05587 to KS).

References

Abiko T, Miyasaka SC (2020) Aerenchyma and barrier to radial oxygen loss are formed in roots of Taro (*Colocasia esculenta*) propagules under flooded conditions. J Plant Res 133(1):49–56

Abiko T, Kotula L, Shiono K, Malik AI, Colmer TD, Nakazono M (2012) Enhanced formation of aerenchyma and induction of a barrier to radial oxygen loss in adventitious roots of *Zea nicaraguensis* contribute to its waterlogging tolerance as compared with maize (*Zea mays* ssp. mays). Plant Cell Environ 35(9):1618–1630

Aloni R, Enstone DE, Peterson CA (1998) Indirect evidence for bulk water flow in root cortical cell walls of three dicotyledonous species. Planta 207(1):1–7

Aoki D, Yamaguchi H (2008) Genetic relationship between *Echinochloa crus-galli* and *Echinochloa oryzicola* accessions inferred from internal transcribed spacer and chloroplast DNA sequences. Weed Biol Manag 8(4):233–242

Aoki D, Yamaguchi H (2009) Oryza sh4 gene homologue represents homoeologous genomic copies in polyploid Echinochloa. Weed Biol Manag 9(3):225–233

Armstrong W (1970) Rhizosphere oxidation in rice and other species: a mathematical model based on the oxygen flux component. Physiol Plant 23(3):623–630

Armstrong W (1971) Radial oxygen losses from intact rice roots as affected by distance from the apex, respiration and waterlogging. Physiol Plant 25:192–197

Armstrong W (1979) Aeration in higher plants. Adv Bot Res 7:225–332

Armstrong J, Armstrong W (2001) Rice and Phragmites: effects of organic acids on growth, root permeability, and radial oxygen loss to the rhizosphere. Am J Bot 88(8):1359–1370

Armstrong J, Armstrong W (2005) Rice: sulfide-induced barriers to root radial oxygen loss, Fe^{2+} and water uptake, and lateral root emergence. Ann Bot 96(4):625–638

Armstrong W, Beckett PM (1987) Internal aeration and the development of stelar anoxia in submerged roots—a multishelled mathematical-model combining axial diffusion of oxygen in the cortex with radial losses to the stele, the wall layers and the rhizosphere. New Phytol 105(2):221–245

Armstrong J, Armstrong W, Beckett P, Halder J, Lythe S, Holt R, Sinclair A (1996) Pathways of aeration and the mechanisms and beneficial effects of humidity- and venturi-induced convections in *Phragmites australis* (Cav.) Trin. ex Steud. Aquat Bot 54:177–197

Armstrong W, Cousins D, Armstrong J, Turner DW, Beckett PM (2000) Oxygen distribution in wetland plant roots and permeability barriers to gas-exchange with the rhizosphere: a microelectrode and modelling study with *Phragmites australis*. Ann Bot 86(3):687–703

Bailey-Serres J, Voesenek LACJ (2008) Flooding stress: acclimations and genetic diversity. Annu Rev Plant Biol 59:313–339

Bailey-Serres J, Voesenek LA (2010) Life in the balance: a signaling network controlling survival of flooding. Curr Opin Plant Biol 13(5):489–494

Barros J, Serk H, Granlund I, Pesquet E (2015) The cell biology of lignification in higher plants. Ann Bot 115(7):1053–1074

Beroyz G, Casale R, Farreiros A, Palermo M, Margulies M, Voto L, Fabregues G, Ramalingam R, Davies T, Bryce R et al (1994) Clasp—a randomized trial of low-dose aspirin for the prevention and treatment of preeclampsia among 9364 pregnant-women. Lancet 343(8898):619–629

Bertholdsson N-O (2013) Screening for barley waterlogging tolerance in Nordic barley cultivars (*Hordeum vulgare* L.) using chlorophyll fluorescence on hydroponically-grown plants. Agronomy 3(2):376

Blossfeld S, Gansert D, Thiele B, Kuhn AJ, Lösch R (2011) The dynamics of oxygen concentration, pH value, and organic acids in the rhizosphere of Juncus spp. Soil Biol Biochem 43(6):1186–1197

Bücker-Neto L, Paiva ALS, Machado RD, Arenhart RA, Margis-Pinheiro M (2017) Interactions between plant hormones and heavy metals responses. Genet Mol Biol 40(1 suppl 1):373–386

Campbell MM, Sederoff RR (1996) Variation in lignin content and composition—mechanisms of control and implications for the genetic improvement of plants. Plant Physiol 110(1):3–13

Cheng H, Chen D-T, Tam NF-Y, Chen G-Z, Li S-Y, Ye Z-H (2012) Interactions among Fe^{2+}, S^{2-}, and Zn^{2+} tolerance, root anatomy, and radial oxygen loss in mangrove plants. J Exp Bot 63:2619–2630

Colmer TD (2003a) Aerenchyma and an inducible barrier to radial oxygen loss facilitate root aeration in upland, paddy and deep-water rice (*Oryza sativa* L.). Ann Bot 91(2):301–309

Colmer TD (2003b) Long-distance transport of gases in plants: a perspective on internal aeration and radial oxygen loss from roots. Plant Cell Environ 26(1):17–36

Colmer TD, Voesenek LACJ (2009) Flooding tolerance: suites of plant traits in variable environments. Funct Plant Biol 36(8):665–681

Colmer TD, Gibberd MR, Wiengweera A, Tinh TK (1998) The barrier to radial oxygen loss from roots of rice (*Oryza sativa* L.) is induced by growth in stagnant solution. J Exp Bot 49(325):1431–1436

Colmer TD, Cox MCH, Voesenek LACJ (2006) Root aeration in rice (*Oryza sativa*): evaluation of oxygen, carbon dioxide, and ethylene as possible regulators of root acclimatizations. New Phytol 170(4):767–777

Colmer TD, Kotula L, Malik AI, Takahashi H, Konnerup D, Nakazono M, Pedersen O (2019) Rice acclimation to soil flooding: low concentrations of organic acids can trigger a barrier to radial oxygen loss in roots. Plant Cell Environ 42(7):2183–2197

Connell EL, Colmer TD, Walker DI (1999) Radial oxygen loss from intact roots of *Halophila ovalis* as a function of distance behind the root tip and shoot illumination. Aquat Bot 63(3–4):219–228

Davison W, Fones GR, Grime GW (1997) Dissolved metals in surface sediment and a microbial mat at 100-μm resolution. Nature 387(6636):885–888

De Simone O, Haase K, Müller E, Junk WJ, Gonsior G, Schmidt W (2002) Impact of root morphology on metabolism and oxygen distribution in roots and rhizosphere from two Central Amazon floodplain tree species. Funct Plant Biol 29(9):1025–1035

De Simone O, Haase K, Muller E, Junk WJ, Hartmann K, Schreiber L, Schmidt W (2003) Apoplasmic barriers and oxygen transport properties of hypodermal cell walls in roots from four Amazonian tree species. Plant Physiol 132(1):206–217

Denison RF (1998) Decreased oxygen permeability: a universal stress response in legume root nodules. Bot Acta 111(3):191–192

Denison RF, Hunt S, Layzell DB (1992) Nitrogenase activity, nodule respiration, and O_2 permeability following detopping of alfalfa and birdsfoot trefoil. Plant Physiol 98(3):894–900

Drapek C, Sparks EE, Marhavy P, Taylor I, Andersen TG, Hennacy JH, Geldner N, Benfey PN (2018) Minimum requirements for changing and maintaining endodermis cell identity in the Arabidopsis root. Nat Plants 4(8):586–595

Ejiri M, Shiono K (2019) Prevention of radial oxygen loss is associated with exodermal suberin along adventitious roots of annual wild species of Echinochloa. Front Plant Sci 10:254

Ejiri M, Shiono K (2020) Groups of multi-cellular passage cells in the root exodermis of *Echinochloa crus-galli* varieties lack not only suberin lamellae but also lignin deposits. Plant Signal Behav 15(2):1719749

Ejiri M, Shiono K (2023) Characteristic of morphology, anatomy and histology of a seminal root of wild rice, *Oryza glumaepatula*. Root Res 32(1):3–14

Ejiri M, Sawazaki Y, Shiono K (2020) Some accessions of Amazonian wild rice (*Oryza glumaepatula*) constitutively form a barrier to radial oxygen loss along adventitious roots under aerated conditions. Plant Theory 9(7):880

Ejiri M, Fukao T, Miyashita T, Shiono K (2021) A barrier to radial oxygen loss helps the root system cope with waterlogging-induced hypoxia. Breed Sci 71(1):40–50

Enstone DE, Peterson CA (1997) Suberin deposition and band plasmolysis in the corn (*Zea mays* L.) root exodermis. Can J Bot 75(7):1188–1199

Enstone DE, Peterson CA (2005) Suberin lamella development in maize seedling roots grown in aerated and stagnant conditions. Plant Cell Environ 28(4):444–455

Enstone DE, Peterson CA, Ma F (2003) Root endodermis and exodermis: structure, function, and responses to the environment. J Plant Growth Regul 21(4):335–351

Ernst WHO (1990) Ecophysiology of plants in waterlogged and flooded environments. Aquat Bot 38(1):73–90

Fukao T, Barrera-Figueroa BE, Juntawong P, Peña-Castro JM (2019) Submergence and waterlogging stress in plants: a review highlighting research opportunities and understudied aspects. Front Plant Sci 10:340

Garthwaite AJ, von Bothmer R, Colmer TD (2003) Diversity in root aeration traits associated with waterlogging tolerance in the genus Hordeum. Funct Plant Biol 30(8):875–889

Garthwaite AJ, Armstrong W, Colmer TD (2008) Assessment of O_2 diffusivity across the barrier to radial O_2 loss in adventitious roots of *Hordeum marinum*. New Phytol 179(2):405–416

Graça J (2015) Suberin: the biopolyester at the frontier of plants. Front Chem 3:62

Healy MT, Armstrong W (1972) The effectiveness of internal oxygen transport in a mesophyte (*Pisum sativum* L.). Planta 103(4):302–309

Hirabayashi Y, Mahendran R, Koirala S, Konoshima L, Yamazaki D, Watanabe S, Kim H, Kanae S (2013) Global flood risk under climate change. Nat Clim Change 3(9):816–821

Hirabayashi Y, Tanoue M, Sasaki O, Zhou X, Yamazaki D (2021) Global exposure to flooding from the new CMIP6 climate model projections. Sci Rep 11(1):3740

Jackson MB, Fenning TM, Drew MC, Saker LR (1985) Stimulation of ethylene production and gas-space (aerenchyma) formation in adventitious roots of *Zea mays* L. by small partial pressures of oxygen. Planta 165(4):486–492

James EK, Sprent JI, Minchin FR, Brewin NJ (1991) Intercellular location of glycoprotein in soybean nodules: effect of altered rhizosphere oxygen concentration. Plant Cell Environ 14(5):467–476

Jiménez JC, Kotula L, Veneklaas EJ, Colmer TD (2019) Root-zone hypoxia reduces growth of the tropical forage grass *Urochloa humidicola* in high-nutrient but not low-nutrient conditions. Ann Bot 124(6):1019–1032

Jiménez JC, Clode PL, Signorelli S, Veneklaas EJ, Colmer TD, Kotula L (2021a) The barrier to radial oxygen loss impedes the apoplastic entry of Fe into the roots of *Urochloa humidicola*. J Exp Bot 72:3279

Jiménez JC, Pellegrini E, Pedersen O, Nakazono M (2021b) Radial oxygen loss from plant roots—methods. Plant Theory 10(11):2322

Justin SHFW, Armstrong W (1987) The anatomical characteristics of roots and plant response to soil flooding. New Phytol 106:465–495

Konnerup D, All M, Islam AKMR, Colmer TD (2017) Evaluation of root porosity and radial oxygen loss of disomic addition lines of *Hordeum marinum* in wheat. Funct Plant Biol 44(4):400–409

Kotula L, Ranathunge K, Schreiber L, Steudle E (2009) Functional and chemical comparison of apoplastic barriers to radial oxygen loss in roots of rice (*Oryza sativa* L.) grown in aerated or deoxygenated solution. J Exp Bot 60(7):2155–2167

Kotula L, Colmer TD, Nakazono M (2014) Effects of organic acids on the formation of the barrier to radial oxygen loss in roots of *Hordeum marinum*. Funct Plant Biol 41(2):187–202

Kreszies T, Schreiber L, Ranathunge K (2018) Suberized transport barriers in Arabidopsis, barley and rice roots: from the model plant to crop species. J Plant Physiol 227:75–83

Kreszies T, Eggels S, Kreszies V, Osthoff A, Shellakkutti N, Baldauf JA, Zeisler-Diehl VV, Hochholdinger F, Ranathunge K, Schreiber L (2020) Seminal roots of wild and cultivated barley differentially respond to osmotic stress in gene expression, suberization, and hydraulic conductivity. Plant Cell Environ 43(2):344–357

Kreuzwieser J, Papadopoulou E, Rennenberg H (2004) Interaction of flooding with carbon metabolism of forest trees. Plant Biol 6(3):299–306

Kulichikhin K, Yamauchi T, Watanabe K, Nakazono M (2014) Biochemical and molecular characterization of rice (*Oryza sativa* L.) roots forming a barrier to radial oxygen loss. Plant Cell Environ 37(10):2406–2420

Laanbroek HJ (1990) Bacterial cycling of minerals that affect plant-growth in waterlogged soils—a review. Aquat Bot 38(1):109–125

Lamers LPM, Tomassen HBM, Roelofs JGM (1998) Sulfate-induced entrophication and phytotoxicity in freshwater wetlands. Environ Sci Technol 32(2):199–205

Larsen M, Santner J, Oburger E, Wenzel W, Glud R (2015) O_2 dynamics in the rhizosphere of young rice plants (*Oryza sativa* L.) as studied by planar optodes. Plant Soil 390(1–2):279–292

Lenzewski N, Mueller P, Meier RJ, Liebsch G, Jensen K, Koop-Jakobsen K (2018) Dynamics of oxygen and carbon dioxide in rhizospheres of *Lobelia dortmanna*—a planar optode study of belowground gas exchange between plants and sediment. New Phytol 218(1):131–141

Li P, Yu Q, Gu X, Xu C, Qi S, Wang H, Zhong F, Baskin TI, Rahman A, Wu S (2018) Construction of a functional Casparian strip in non-endodermal lineages is orchestrated by two parallel signaling systems in *Arabidopsis thaliana*. Curr Biol 28(17):2777–2786

Lu H-P, Gao Q, Han J-P, Guo X-H, Wang Q, Altosaar I, Barberon M, Liu J-X, Gatehouse AMR, Shu Q-Y (2022) An ABA-serotonin module regulates root suberization and salinity tolerance. New Phytol 236(3):958–973

Maisch M, Lueder U, Kappler A, Schmidt C (2019) Iron lung: how rice roots induce iron redox changes in the rhizosphere and create niches for microaerophilic Fe(II)-oxidizing bacteria. Environ Sci Technol Lett 6(10):600–605

Malik AI, Islam AKMR, Colmer TD (2011) Transfer of the barrier to radial oxygen loss in roots of *Hordeum marinum* to wheat (*Triticum aestivum*): evaluation of four *H. marinum*–wheat amphiploids. New Phytol 190(2):499–508

Mano Y, Nakazono M (2021) Genetic regulation of root traits for soil flooding tolerance in genus Zea. Breed Sci 71(1):30–39

Manzur ME, Grimoldi AA, Insausti P, Striker GG (2015) Radial oxygen loss and physical barriers in relation to root tissue age in species with different types of aerenchyma. Funct Plant Biol 42(1):9–17

Martin BC, Bougoure J, Ryan MH, Bennett WW, Colmer TD, Joyce NK, Olsen YS, Kendrick GA (2019) Oxygen loss from seagrass roots coincides with colonisation of sulphide-oxidising cable bacteria and reduces sulphide stress. ISME J 13(3):707–719

McDonald MP, Galwey NW, Colmer TD (2001a) Waterlogging tolerance in the tribe Triticeae: the adventitious roots of *Critesion marinum* have a relatively high porosity and a barrier to radial oxygen loss. Plant Cell Environ 24(6):585–596

McDonald MP, Galwey NW, Ellneskog-Staam P, Colmer TD (2001b) Evaluation of *Lophopyrum elongatum* as a source of genetic diversity to increase the waterlogging tolerance of hexaploid wheat (*Triticum aestivum*). New Phytol 151(2):369–380

McDonald MP, Galwey NW, Colmer TD (2002) Similarity and diversity in adventitious root anatomy as related to root aeration among a range of wetland and dryland grass species. Plant Cell Environ 25(3):441–451

Mongon J, Konnerup D, Colmer TD, Rerkasem B (2014) Responses of rice to Fe^{2+} in aerated and stagnant conditions: growth, root porosity and radial oxygen loss barrier. Funct Plant Biol 41(9):922–929

Neubauer SC, Toledo-Duran GE, Emerson D, Megonigal JP (2007) Returning to their roots: iron-oxidizing bacteria enhance short-term plaque formation in the wetland-plant rhizosphere. Geomicrobiol J 24(1):65–73

Nishiuchi S, Yamauchi T, Takahashi H, Kotula L, Nakazono M (2012) Mechanisms for coping with submergence and waterlogging in rice. Rice 5(1):2

Noorrohmah S, Takahashi H, Nakazono M (2020) Formation of a barrier to radial oxygen loss in L-type lateral roots of rice. Plant Root 14:33–41

Parsons R, Day DA (1990) Mechanism of soybean nodule adaptation to different oxygen pressures. Plant Cell Environ 13(6):501–512
Pedersen O, Perata P, Voesenek LACJ (2017) Flooding and low oxygen responses in plants. Funct Plant Biol 44(9):3–6
Pedersen O, Sauter M, Colmer TD, Nakazono M (2021) Regulation of root adaptive anatomical and morphological traits during low soil oxygen. New Phytol 229(1):42–49
Peralta Ogorek LL, Pellegrini E, Pedersen O (2021) Novel functions of the root barrier to radial oxygen loss—radial diffusion resistance to H and water vapour. New Phytol 231(4):1365–1376
Peralta Ogorek LL, Takahashi H, Nakazono M, Pedersen O (2023) The barrier to radial oxygen loss protects roots against hydrogen sulphide intrusion and its toxic effect. New Phytol 238(5):1825–1837
Peterson CA, Enstone DE (1996) Functions of passage cells in the endodermis and exodermis of roots. Physiol Plant 97(3):592–598
Ponnamperuma FN (1984) Effects of flooding on soils. In: Kozlowski TT (ed) Flooding and plant growth. Academic Press, New York, pp 9–45
Ranathunge K, Lin J, Steudle E, Schreiber L (2011a) Stagnant deoxygenated growth enhances root suberization and lignifications, but differentially affects water and NaCl permeabilities in rice (*Oryza sativa* L.) roots. Plant Cell Environ 34(8):1223–1240
Ranathunge K, Schreiber L, Franke R (2011b) Suberin research in the genomics era—new interest for an old polymer. Plant Sci 180(3):399–413
Ranathunge K, Schreiber L, Bi YM, Rothstein SJ (2016) Ammonium-induced architectural and anatomical changes with altered suberin and lignin levels significantly change water and solute permeabilities of rice (*Oryza sativa* L.) roots. Planta 243(1):231–249
Rao AN, Johnson DE, Sivaprasad B, Ladha JK, Mortimer AM (2007) Weed management in direct-seeded rice. In: Sparks DL (ed) Advances in agronomy. Academic Press, pp 153–255
Schreiber L, Franke RB (2011) Endodermis and exodermis in roots. Wiley, Chichester, pp 1–7
Schreiber L, Hartmann K, Skrabs M, Zeier J (1999) Apoplastic barriers in roots: chemical composition of endodermal and hypodermal cell walls. J Exp Bot 50(337):1267–1280
Seago JL, Marsh LC, Stevens KJ, Soukup A, Votrubová O, Enstone DE (2005) A re-examination of the root cortex in wetland flowering plants with respect to aerenchyma. Ann Bot 96(4):565–579
Serra O, Geldner N (2022) The making of suberin. New Phytol 235(3):848–866
Setter TL, Waters I (2003) Review of prospects for germplasm improvement for waterlogging tolerance in wheat, barley and oats. Plant Soil 253(1):1–34
Shiono K, Matsuura H (2024) Exogenous abscisic acid induces the formation of a suberized barrier to radial oxygen loss in adventitious roots of barley (*Hordeum vulgare*). Ann Bot. In Press
Shiono K, Yamada S (2014) Waterlogging tolerance and capacity for oxygen transport in *Brachypodium distachyon* (Bd21). Plant Root 8:5–12
Shiono K, Takahashi H, Colmer TD, Nakazono M (2008) Role of ethylene in acclimations to promote oxygen transport in roots of plants in waterlogged soils. Plant Sci 175(1–2):52–58
Shiono K, Ogawa S, Yamazaki S, Isoda H, Fujimura T, Nakazono M, Colmer TD (2011) Contrasting dynamics of radial O_2-loss barrier induction and aerenchyma formation in rice roots of two lengths. Ann Bot 107(1):89–99
Shiono K, Yamauchi T, Yamazaki S, Mohanty B, Malik AI, Nagamura Y, Nishizawa NK, Tsutsumi N, Colmer TD, Nakazono M (2014) Microarray analysis of laser-microdissected tissues indicates the biosynthesis of suberin in the outer part of roots during formation of a barrier to radial oxygen loss in rice (*Oryza sativa*). J Exp Bot 65(17):4795–4806
Shiono K, Ejiri M, Shimizu K, Yamada S (2019) Improved waterlogging tolerance of barley (*Hordeum vulgare*) by pretreatment with ethephon. Plant Prod Sci 22(2):285–295
Shiono K, Koshide A, Iwasaki K, Oguri K, Fukao T, Larsen M, Glud RN (2022a) Imaging the snorkel effect during submerged germination in rice: oxygen supply via the coleoptile triggers seminal root emergence underwater. Front Plant Sci 13:946776

Shiono K, Yoshikawa M, Kreszies T, Yamada S, Hojo Y, Matsuura T, Mori IC, Schreiber L, Yoshioka T (2022b) Abscisic acid is required for exodermal suberization to form a barrier to radial oxygen loss in the adventitious roots of rice (*Oryza sativa*). New Phytol 233(2):655–669

Smits AJM, Laan P, Thier RH, van der Velde G (1990) Root aerenchyma, oxygen leakage patterns and alcoholic fermentation ability of the roots of some nymphaeid and isoetid macrophytes in relation to the sediment type of their habitat. Aquat Bot 38(1):3–17

Sorrell BK, Mendelssohn IA, McKee KL, Woods RA (2000) Ecophysiology of wetland plant roots: a modelling comparison of aeration in relation to species distribution. Ann Bot 86(3):675–685

Soukup A, Armstrong W, Schreiber L, Franke R, Votrubová O (2007) Apoplastic barriers to radial oxygen loss and solute penetration: a chemical and functional comparison of the exodermis of two wetland species, *Phragmites australis* and *Glyceria maxima*. New Phytol 173(2):264–278

Sun L, Di D-W, Li G, Kronzucker HJ, Wu X, Shi W (2020) Endogenous ABA alleviates rice ammonium toxicity by reducing ROS and free ammonium via regulation of the SAPK9–bZIP20 pathway. J Exp Bot 71(15):4562–4577

Tanesaka E, Ohno T, Yamaguchi H (2010) Species diversity of the genus Echinochloa (Poaceae), native to eastern Australia a focus on their habitat and the threat of exotic species. J Crop Res 55:13–17

Visser EJW, Colmer TD, Blom CWPM, Voesenek LACJ (2000) Changes in growth, porosity, and radial oxygen loss from adventitious roots of selected mono- and dicotyledonous wetland species with contrasting types of aerenchyma. Plant Cell Environ 23(11):1237–1245

Voesenek LACJ, Sasidharan R (2013) Ethylene—and oxygen signalling—drive plant survival during flooding. Plant Biol 15(3):426–435

Voesenek LACJ, Armstrong W, Bogemann GM, McDonald MP, Colmer TD (1999) A lack of aerenchyma and high rates of radial oxygen loss from the root base contribute to the waterlogging intolerance of *Brassica napus*. Aust J Plant Physiol 26(1):87–93

Wang C, Wang H, Li P, Li H, Xu C, Cohen H, Aharoni A, Wu S (2020) Developmental programs interact with abscisic acid to coordinate root suberization in Arabidopsis. Plant J 104(1):241–251

Watanabe K, Nishiuchi S, Kulichikhin K, Nakazono M (2013) Does suberin accumulation in plant roots contribute to waterlogging tolerance? Front Plant Sci 4:178

Watanabe K, Takahashi H, Sato S, Nishiuchi S, Omori F, Malik AI, Colmer TD, Mano Y, Nakazono M (2017) A major locus involved in the formation of the radial oxygen loss barrier in adventitious roots of teosinte *Zea nicaraguensis* is located on the short-arm of chromosome 3. Plant Cell Environ 40(2):304–316

Yamasue Y (2001) Strategy of *Echinochloa oryzicola* Vasing. for survival in flooded rice. Weed Biol Manag 1(1):28–36

Yamauchi T, Colmer TD, Pedersen O, Nakazono M (2018) Regulation of root traits for internal aeration and tolerance to soil waterlogging-flooding stress. Plant Physiol 176(2):1118–1130

Yamauchi T, Abe F, Tsutsumi N, Nakazono M (2019) Root cortex provides a venue for gas-space formation and is essential for plant adaptation to waterlogging. Front Plant Sci 10:259

Chapter 9
Oxygen Transport and Plant Ventilation

Gustavo G. Striker

9.1 Introduction

Internal transport of gases is critical for plants inhabiting natural or crop lands prone to experience flooding of various intensities. The diffusivity of gases in water is between 5000 and 10,000 times slower than in air; thereby, the exchange of gases between submerged tissues and the surrounding water environment is severely restricted (Armstrong 1979). So, aerobic respiration shifts toward fermentation, and the production of energy (ATP) drops rapidly. In this scenario, the energy crisis is evident, and the fermentation often derives in tissues accumulation of metabolically toxic products such as ethanol. Interestingly, the adaptation to waterlogging/submergence-induced hypoxia is not based on a more efficient use of oxygen by the tolerant species. In fact, the critical oxygen pressure for respiration (COPR) is quite similar among species with different tolerance but phylogenetically related (e.g., species of the same genus; see Armstrong et al. 2009). So, the tolerance is related to a better supply of oxygen to the cells rather than the efficiency on the use of molecular oxygen. To achieve a proper supply of oxygen to the cells, plants develop anatomical adaptations so that the transport of gases is facilitated. The most conspicuous response to root-zone oxygen deficiency is the formation of aerenchyma in the root cortex (Justin and Armstrong 1987; Visser et al. 2000; Colmer 2003; Striker et al. 2007; Manzur et al. 2014; Yamauchi and Nakazono 2022). This tissue is composed by longitudinally interconnected gas spaces, which provide a low-resistance internal pathway for gas transport (Armstrong 1979; Colmer 2003). By using this

G. G. Striker (✉)
IFEVA, Universidad de Buenos Aires, CONICET, Facultad de Agronomía, Buenos Aires, Argentina

School of Agriculture and Environment, Faculty of Science, The University of Western Australia, Perth, WA, Australia
e-mail: striker@agro.uba.ar

J.-I. Sakagami, M. Nakazono (eds.), *Responses of Plants to Soil Flooding*,
https://doi.org/10.1007/978-981-99-9112-9_9

pathway, the oxygen moves to the roots (and the rhizosphere), while carbon dioxide, ethylene, and methane are released from the soil to the shoots and the atmosphere (Armstrong et al. 1996a; Yin et al. 2021). The gas movement within plant tissues is generally explained by diffusion, but other mechanisms involving pressure-flow ventilation (e.g., convection in stems and rhizomes of several emergent and floating-leaved wetland species; Armstrong et al. 1996a, b; Colmer 2003) are also possible. In the following sections, some basic aspects of aerenchyma formation and other root adaptations related to the transport of gases, as well as the most important mechanisms of oxygen transport will be discussed. At this initial point, it is explicitly declared that the diffusive and pressure-flow ventilation through the gas-filled space continuum in plants has been reviewed in detail in earlier works (e.g., Armstrong et al. 1996a, b; Grosse 1996; Colmer 2003; Wegner 2010; Armstrong and Armstrong 2014; Striker 2023). Therefore, some repetition is inevitable, but the intention of this chapter is to provide the fundamental concepts of each type of mechanism, so the non-specialized reader can have a summarized version of these modes of gas transport without excessive technicalities.

9.2 The Diffusion Mechanism Explains Root Oxygenation

Gas movement into, and along, plant roots is due to diffusion (Armstrong 1979; Colmer 2003). The capacity for longitudinal oxygen diffusion in roots is determined by anatomical, morphological, and physiological traits, but it is also influenced by environmental conditions such as temperature and oxygen demand in the rhizosphere (Colmer 2003). Aerenchyma formation is the most conspicuous response facilitating root aeration (Armstrong 1979; Fig. 9.1a–h). Two main types of (primary) aerenchyma have been recognized: schizogenous and lysigenous aerenchyma (Jackson and Armstrong 1999; Fig. 9.1a, b). Schizogenous aerenchyma forms gas spaces because of cell separation and expansion without cell death. By contrast, lysigenous aerenchyma is developed by the death and subsequent lysis of some cortical cells (Yamauchi et al. 2013). The maximum length of a root growing in an oxygen deficient soil is determined by the internal diffusion of oxygen reaching the apex. Therefore, the knowledge of the factors determining the apex oxygenation is critical.

Anatomical characteristics related to root tissues aeration are (1) the number and size of the aerenchyma lacunae (Fig. 9.1f–h), (2) the packing of cells within the root cortex (Fig. 9.1c–e), (3) the ratio cortex to stele and cortex area, and (4) presence of a barrier to the radial oxygen loss (Justin and Armstrong 1987; Colmer 2003; Striker et al. 2007; Yamauchi et al. 2019, 2021; Fig. 9.1). As waterlogging duration increases, the aerenchyma lacunae become larger and more numerous (Smirnoff and Crawford 1983) in response to (1), firstly, the increase of ethylene (gaseous hormone) in the roots due to physical entrapment in the aqueous media and enhanced expression of the ethylene biosynthesis enzymes and (2), secondly, a burst of reactive oxygen species (ROS) due to the expression of the gene encoding respiratory

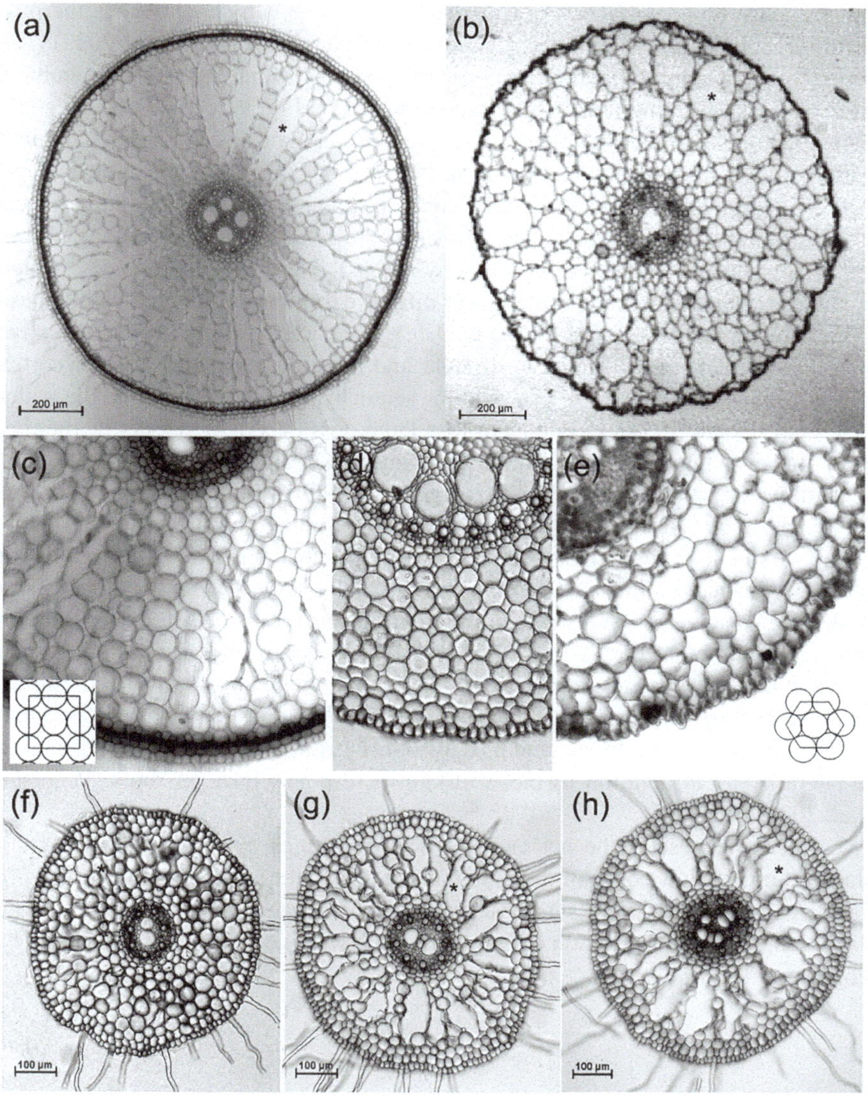

Fig. 9.1 Root cross sections showing lysigenous aerenchyma in *Oryza sativa* (**a**), schizogenous aerenchyma in *Rumex crispus* (**b**), cubic (radial) cell arrangement in all cortex of *O. sativa* (**c**), hexagonal cell arrangement in the outer cortex, and cubic cell arrangement in middle and inner cortex of *Zea mays* (**d**), hexagonal cell arrangement in all cortex of *Ornithopus compressus* (**e**), progression of aerenchyma formation in *Hordeum marinum* at 2 cm (**f**), 4 cm (**g**), and 6 cm (**h**) from the root tip. Asterisks indicate aerenchyma (**a**, **b**, **f**–**h**). (Most images used were generously provided by Lukasz Kotula and Tim Colmer)

burst oxidase homolog (RBOH), triggered by ethylene, which stimulates the programmed cell death (PCD) in the root cortex (Yamauchi and Nakazono 2022). The arrangement of cells within the cortex can follow two general types, namely, cubical and hexagonal packing (Justin and Armstrong 1987; Fig. 9.1c–e). The hexagonal packing is mostly found in upland species, and it is a type of cell arrangement that seldom derives in aerenchyma formation (Fig. 9.1e). Conversely, the cubical packing is a pre-condition related to aerenchyma formation, and it is found in intermediate and wetland species (Justin and Armstrong 1987). The cubical packing allows for the presence of bigger (non-aerenchymatous) intercellular air spaces, which might contribute to facilitate some sort of diffusive transport of oxygen within these tissues (Justin and Armstrong 1987, Fig. 9.1c). The cortex-to-stele ratio is indicative of how well plants are adapted to drought and flooding because the wider cortex (and narrower stele) allows higher area for aerenchyma formation, while the wider stele (and narrower cortex) is related to a large area of xylem vessels facilitating water transport under drought conditions (Yamauchi et al. 2021). The formation (or constitutive presence) of a barrier to oxygen loss (ROL) in the outer cortex due to suberin deposition restricts the oxygen diffusion from the root surfaces to the rhizosphere, so that longitudinal oxygen movement through aerenchyma toward the root apex is enhanced (Colmer 2003; Kotula et al. 2009; Shiono et al. 2014; Yamauchi et al. 2021). As some species not forming barrier to ROL can also develop suberin lamellae in the outer cortex (Abiko et al. 2012), it was proposed that the oxygen permeability may be determined by the suberin monomer composition (or the structure) and not only its presence (Watanabe et al. 2017). Recently, it was revealed that the formation of aerenchyma, and a tight ROL barrier, also improves oxygenation of lateral roots in addition to the well-known response in main roots (Pedersen et al. 2021).

Diffusion represents the net movement of molecules by random motion from places of high to low concentrations so that the concentration gradient is the driving force. A simple equation of steady-state diffusion between an oxygen source and a sink—where the concentration of O_2 is zero—which are connected by a tube of "L" length is as follows:

$$J = D \cdot A \cdot \frac{\Delta C}{L}, \tag{9.1}$$

where J is the diffusion rate, D is the coefficient of diffusion, A is the cross-sectional area of the source, sink, and the tube connecting these, and $\frac{\Delta C}{L}$ is the gradient of concentration for the distance (L) between the oxygen source and sink (see Armstrong and Armstrong 2014). The diffusion coefficient is affected by the mass, temperature, and the characteristics of the medium where the molecules are moving. The mean experimental value of D at 20 °C in air is 0.2 $cm^2\ s^{-1}$, and in pure water, it is about $2 \times 10^{-5}\ cm^2\ s^{-1}$, being sensitive to temperature. The estimation of oxygen diffusion coefficient in aqueous solution can be calculated following the Stokes–Einstein equation (i.e., required parameters are molecular weight of water, the absolute temperature, the solution viscosity, and the molar volume of water). In addition to the high reduction in the diffusion of oxygen when in water media, the concentration of this gas is also highly reduced in water in a range between 1/25 to 1/30, depending on the temperature.

The diffusion of gases in plants does not work as easily as the analogy with the source and sink of oxygen connected by an open tube. Pioneer works by Armstrong have shown with mathematical modeling and biological measurements that several factors must be considered (Armstrong 1979 and following work). The porosity of tissues (i.e., "cross sectional area of the tube") varies with the proportion of aerenchyma, the relative areas of cortex and non-aerenchymatous stele, root thickness, and cortical cell arrangement. The tortuosity of the path along the root (axial movement of oxygen) is little according to Armstrong (1979), but the tortuosity in radial direction (from the outer root cortex toward the rhizosphere or from inner cortex toward the stele) is expected to be of higher magnitude (Armstrong and Armstrong 2014). Literature has informed few values for the tortuosity factor of the diffusive path in roots. Jensen et al. (1967) reported values of ca. 0.44 for rice, barley, and maize. Greenwood (1967) found low values in seedlings of 12 species, while Gaynard (1979) and Armstrong et al. (1982) found no significant tortuosity (values of 1) for wetland species and pea roots, respectively.

The diffusion of oxygen is critical in determining the length of roots growing in anaerobic substrates. Some estimations were made based on calculations of the theoretical maximum root length (*L*) proposed by the model of Armstrong (1979), which is as follows:

$$L = \sqrt{\frac{\left[2\left(D \cdot \tau \cdot P\left(C0 - C1\right)\right)\right]}{M}}, \tag{9.2}$$

where *D* is the oxygen diffusivity as stated earlier, τ represents the tortuosity (assumed to be 1 according to Armstrong et al. 1982), *P* is the root fractional porosity, *C*0 is the oxygen concentration in air (2.72 g cm^{-3} at 20 °C), *C*1 is the oxygen concentration at the root tip (assumed to be zero when root stops growing), and *M* is the rate of oxygen uptake (i.e., respiration) by the root (g O_2 cm^{-3} s^{-1}). This model is based on the assumptions that tissue porosity and respiration are uniform along the root and that there is no radial loss of oxygen (ROL) from the root to the external medium. Both assumptions are not always satisfied as the porosity usually increases from the tip toward root base (Fig. 9.2; see also Striker et al. 2007) and ROL is very variable among species (Visser et al. 2000; Colmer 2003, Manzur et al. 2014; Pedersen et al. 2021). Nonetheless, this equation offers a useful tool to "benchmark" the potential penetration of roots into anaerobic soils considering differences in root porosity and respiration rates (Gibberd et al. 2001). Recently, Pedersen et al. (2021) modified the version of this model to account for scenarios with and without ROL and to include in the respiration factor *M* a variable fraction of cortex respiration (corrected for declining mass-based respiration as tissue porosity increases) to stelar respiration (considered threefold higher compared to cortex respiration).

An example of diffusive transport of gases is shown in Fig. 9.2 taken from Kotula et al. (2015). Here, the formation of aerenchyma increases the pressure of oxygen in the root cortex (0.4–0.5 kPa) of barley plants grown in deoxygenated stagnant

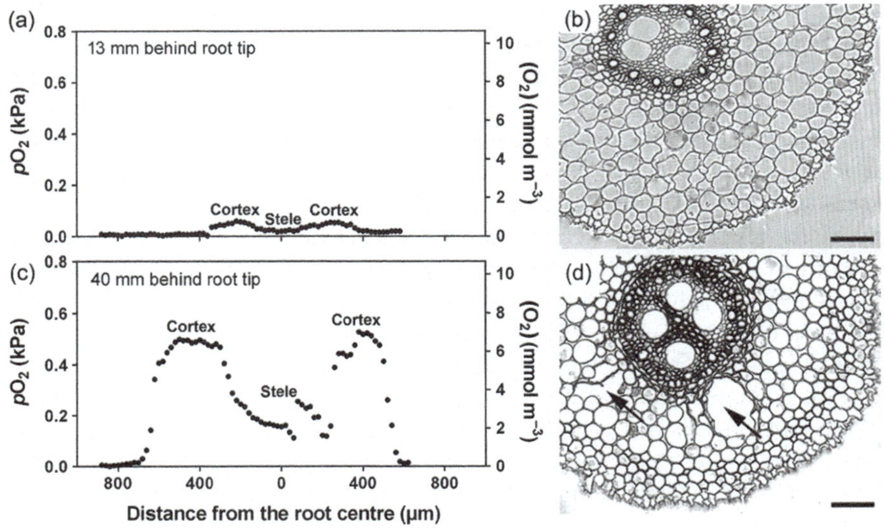

Fig. 9.2 Illustrative radial oxygen profiles showing the importance of aerenchyma formation for diffusive transport of gases taken at apical (**a**) and subapical (**d**) positions along adventitious roots of 70–100-mm-long barley (*Hordeum vulgare*) in deoxygenated stagnant nutrient solution for 14 days. Root cross sections at similar positions are shown in (**b**, 10 mm behind the tip) and (**d**, 50 mm behind the tip) where aerenchyma is indicated by arrows. Bars represent 100 μm. [Adapted from Kotula et al. (2015) with modifications]

nutrient solution (Fig. 9.2c, d). Barley had 19% root porosity, much lower than that attained by rice (>40%; or other waterlogging tolerant species), which can reach higher values of oxygen pressure in the root cortex (>2 kPa when in darkness and >8 kPa when in light; see Pedersen et al. 2009). Still, the range of pressure observed in barley due to aerenchyma formation could satisfy a significant fraction of the respiration requirements according to values modelled by Armstrong and Becket (2011). It should be noticed that aerenchyma is minimum near the root apex where severe hypoxia develops across the entire root radius. In all cases, the stele is particularly hypoxic with very low oxygen pressure (0.045–0.2 kPa) as expected for a tissue of low porosity and high respiratory demand for oxygen (Armstrong et al. 1994; Colmer and Greenway 2011).

9.3 Convection Mechanisms Support the Transport of Oxygen in Stems and Rhizomes of Emergent and Floating-Leaved Species

Pressurized flows of gases require a pressure gradient between two ends of a pathway of low resistance, ending up in an exit to the atmosphere (Beckett et al. 1988). The rates of flow are determined by the magnitude of the pressure gradient and the resistance to flow in the aeration system (Armstrong et al. 1992; Brix et al. 1992). The internal flow of air through the gas-space systems of plants moves the oxygen at much higher speed than the diffusion mechanism commented above. There are at least three mechanisms related to this type of gas movement that will be discussed in the following paragraphs.

9.3.1 Humidity-Induced Pressurization (Positive-Pressure Flows Generated in Living Aerial Tissues)

This type of pressurized flow is generated in living tissues, being the most dominant mode of pressurization occurring in wetland and aquatic plants. It responds to a gradient in water vapor concentration [$\Delta(H_2O_{\text{(vapor)}})$] across microporous partition (<1–3 μm of diameter, allowing Knudsen diffusion), which is more resistant to pressurized backflow than diffusion. The increasing water vapor concentration in gas-filled spaces within the plant (e.g., sub-stomatal cavities and aerenchyma lacunae), which occurs during the process of humidification, dilutes the concentration of other gases like O_2 and N_2. Therefore, a gradient for these gases is created, which now enter via diffusion from the atmosphere (which is drier) through stomata of leaf sheaths (or nodes) of intact culms and increase their pressure within the plant. If there is no outflow path, the static pressure reach follows this equation:

$$\Delta P\text{s} = \left(P\text{atm} + P\text{w inside} - P\text{w atm}\right) - P\text{atm}, \tag{9.3}$$

where ΔPs is the gradient of static pressure, Patm is the atmospheric pressure, Pw inside is the water vapor pressure inside the gas filled space of the plant, and Pw atm is the water vapor pressure of the atmosphere (all units in Pa, following Armstrong et al. 1996a). Pressurization takes place in cases where the partition has micropores up to 3 μm width (Leuning 1983), although overall porosity of the partition, the length of the diffusion path, and temperature are factors influencing the process (Colmer 2003). If there is an "outlet" structure, the pressurized gas will flow following the path of the lowest resistance and being proportional to the pressure difference; this is the "dynamic" pressure, which is lower than the static pressure. The flow usually goes through the shoot/petiole and rhizome and then vents out through porous old plant organs (e.g., dead broken culms; see Fig. 9.3a).

Gradients in temperature can also contribute to humidity-induced pressurization. The air in gas-filled spaces in leaves is saturated with water vapor. The concentration of water vapor inside the leaf depends on leaf temperature where rises in temperature determine marked increases in water vapor concentration. The radiation absorbed by the leaf during daylight determines an increase in leaf temperature and consequently in the water vapor concentration inside the gas-filled spaces of the leaf. This creates a leaf-to-air gradient of water vapor and results in an increasing diffusion of gases (e.g., oxygen and nitrogen) into the leaf, provoking pressurization and mass flow within the plant (Steinberg 1996).

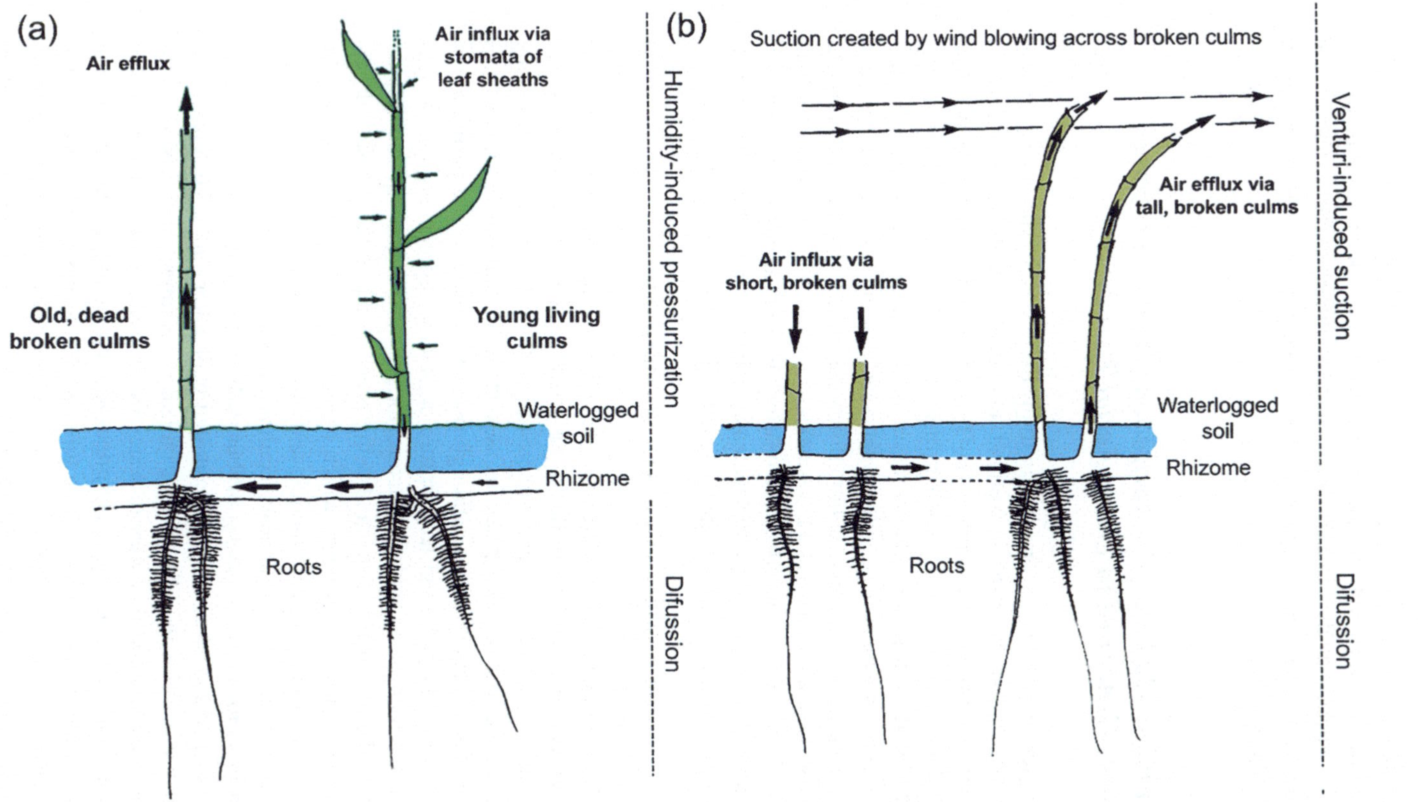

Fig. 9.3 Sketches of humidity-induced pressurization (**a**) and venturi-induced suction (**b**) mechanisms in *Phragmites australis* modified from Armstrong et al. (1996a) (with permission of Elsevier). In (**a**) the inward diffusion of oxygen through leaf sheath stomata is induced by the difference in humidity between the plant's substomatal gas spaces and the atmosphere. In (**b**), a suction pressure is created by the wind when passing across broken culms. In both cases (**a**, **b**), the resulting pressurized gas flow through the rhizome increases the oxygen pressure at the rhizome–root junction, favoring the diffusion of oxygen along the root and to the rhizosphere

Humidity-induced pressurization was identified in *Phragmites australis* (Armstrong and Armstrong 1990; Brix et al. 1992 among others), *P. vallatoria* (Konnerup et al. 2011), *Typha domingensis*, *T. orientalis*, *T. angustifolia*, *T. latifolia*, *Juncus ingens*, *Schoenoplectus validus* (Brix et al. 1992, 1996; Bendix et al. 1994), *Eleocharis sphacelata* (Sorrell et al. 1997), *E. dulcis* (Konnerup et al. 2011), and *Annona squamosa* explants (Armstrong et al. 1997), among others. The static pressure gradient (i.e., flow blocked) and flow rates are provided in Table 9.1, so the reader can become familiar with the values.

9.3.2 Thermal Osmosis

This process, also known as thermal transpiration, involves the gas flow driven by temperature differences across a microporous partition under isobaric conditions (Grosse 1996). The net movement of gas passing through the porous partition is from the cold side, where gas molecules are more concentrated, toward the warm side. So, the direction of the movement occurs against the heat flow (Armstrong et al. 1996b). This process does not involve a gradient in water vapor as in humidity-induced pressurization, and it can even occur in cases when the gases at each side of the partition are dry (Armstrong and Armstrong 2014). The static differential of pressure (i.e., without flow) responds to the following equation in the cases of a partition with very small microporous (diameter <0.1 μm; Armstrong et al. 1996b):

$$\Delta P\text{s} = \left[P\text{atm}\left(T\text{i} / T\text{o} \right)^{1/2} \right] - P\text{atm}, \tag{9.4}$$

where ΔPs is the gradient of static pressure, *P*atm is the atmospheric pressure, and *T*i and *T*o are the temperatures inside and outside the gas-filled space of the plant separated by a microporous partition (pressure units in Pa and temperature in °K). For the flow to occur, there is a need of an "outlet" structure at a certain distance to support the gas movement (i.e., mass flow). It was proposed that tissue layers with small pores inside newly emergent young leaves, which separate the aerenchyma lacunae from the atmosphere, create the pressurized ventilation (influx; Grosse 1996). On the other hand, the mature leaves have more dilated pores (i.e., wider pores), which facilitates the free flow of gas moving out to the atmosphere. So, in aquatic plants, thermal osmosis can operate during the growing season if at least two floating leaves are present.

Thermal osmosis was suggested in some floating-leaved aquatic species, like *Nuphar lutea* and *Nelumbo nucifera*, as the main mechanism of convection/pressurization (Mevi-Schutz and Grosse 1988; Grosse and Bauch 1991); nevertheless, it is important to point out that it is not possible to account for the effects of thermal osmosis and humidity-induced pressurization separately in the "multicomponent gas mixture" of leaves and atmosphere (as defined by Colmer 2003). Moreover, Armstrong and Armstrong (1994) reported only small flows and pressures for

Table 9.1 Gradient of static pressure (Pa) relative to atmospheric pressure developed by humidity-induced pressurization for nine species taken from relevant literature

Species	Measurement conditions	Δ Pressure (Pa) (static)	Flow rate ($cm^3\ min^{-1}$)	References
Typha latifolia	Measures during light and dark periods incubated at 15–25 °C	350	8.0	Bendix et al. (1994)
Typha angustifolia		550	3.5	
Typha domingensis	Measurement of flows through detached culms in February–March in New South Wales	780	3.4	Brix et al. (1992)
Typha orientalis		1070	4.4	
Juncus ingens		222	1.2	
Eleocharis sphacelata		1080	0.85	
	Data modelled to satisfy tissue oxygen demand and prevent anoxia in submerged tissue At 20 °C (4 kPa O_2)	0.81	0.18	Sorrell et al. (1997)
Phragmites australis	Measurements in the early afternoon of summer (Denmark) with air temperature 27 °C, PAR 1000 μmol $m^{-2}\ s^{-1}$ and RH 50%	575 (detached shoots) 576 (shoots connected to stubbles)	4.8 (detached shoots) 2.6 (shoots connected to stubbles)	Brix et al. (1996)
	Measurement of flows through detached culms in February–March in New South Wales	573	5.3	Brix et al. (1992)
	Measurement of leafy culms both in laboratory and field conditions in summer in the UK	Up to 800 (static ΔP in absence of venting) <5 (dynamic ΔP in open culms)	16	Armstrong and Armstrong (1991)
Eleocharis dulcis	Measurement on detached culms or leaf sheaths in the field on sunny days in Vietnam (tropical environment: air temp. 34–36 °C, PAR 1600–1900 μmol $m^{-2}\ s^{-1}$, and RH 47–49)	628	11.9	Konnerup et al. (2011)
Phragmites vallatoria		482	1.59	

Phragmites australis, and Steinberg (1996) stated that values for static pressure only attained up to 30% of theoretical values. By contrast, Dedes and Woermann (1996) reported static pressures close to potential values, experimentally using two types of microporous commercial membranes (i.e., partition) with matrices formed by cellulose nitrate.

9.3.3 Venturi-Induced Suction

In this type of convective mechanism, the wind passes over broken culms creating a suction that moves gases from the culms to the rhizome system; other culms protected from the wind (e.g., given their location within the canopy) are the exit points of such air (see Fig. 9.3b). This suction means that the pressure differential is negative, which contrasts with the positive pressure differential generated by humidity-induced convection and thermal osmosis. The resulting difference in pressure follows the Bernoulli's equation:

$$\Delta P = -\left(\frac{1}{2}\rho \upsilon^2\right), \tag{9.5}$$

where ΔP is the suction pressure (Pa), ρ is the air viscosity (ranging between 1.24 kg m^{-3} at 10 °C to 1.13 kg m^{-3} at 40 °C), and v is the wind speed (m s^{-1}).

Venturi-induced suction was firstly reported by Armstrong et al. (1992) in *Phragmites australis*. It can be seen from the equation that the flow is directly affected by the wind velocity. Pioneer works by Armstrong et al. (1992) stated that this type of pressurized convection can occur in any season and time of day if there is enough wind to start the process. These authors also declared that for the same average wind speed, a gusty wind aerates the rhizomes better than a steady wind. Armstrong et al. (1996a) illustrated this idea proving that the aeration potential of rhizomes can be duplicated by comparing an average steady wind speed of 3 m s^{-1} for 2 min against a gusting wind of 6 m s^{-1} for 1 min, followed by a non-wind situation for another minute. The internal resistance to this flow is determined by the diameter of culms, with wider culms having a lower resistance than narrow ones (Colmer 2003).

Tissue aeration in some species operates through both types of mechanisms, the pressurized flow of gases and diffusion. In wetland plants capable of generating pressurized flows aerating rhizomes, the oxygen movement within roots remains to be diffusive (Beckett et al. 1988). Moreover, the aeration of rhizome tips, buds, and root tips (named blind ends by Armstrong and Armstrong 2014) is also diffusive. The pressurized flows increase the oxygen concentration at the root–rhizome junction compared to what can be expected from only arriving by diffusion. In this regard, it was reported that oxygen increased from 9 to 20% in rhizomes as consequence of pressurization in *Phragmites australis* and that rise in oxygen levels turned into a fourfold increase in the radial oxygen loss by the tip of adventitious roots (Armstrong et al. 1992), which implies an enhanced diffusion of oxygen reaching the tip.

9.4 Venting of Ethylene, CO_2, Methane, and Nitrous Oxide to the Atmosphere Through Plants

The system of gas-filled spaces facilitating the shoot to root oxygen movement also allows for the venting of accumulated gases from submerged tissues toward the atmosphere. Main gases that can be vented through aerenchymatic tissues of plants include ethylene, carbon dioxide, methane, and nitrous oxide, with these latter three being part of the greenhouse effect gases associated to global warming and climate change.

Ethylene is a gaseous hormone that accumulates in root systems due to slow diffusion rate of gases in flooded soil. This hormone triggers numerous responses in root tissues, including lysigenous aerenchyma formation, adventitious rooting, as well as promotion or inhibition of root elongation depending on the ethylene concentration (Sasidharan and Voesenek 2015). If ethylene concentration is high ($>1\ \mu L\ L^{-1}$; Visser et al. 1997), root extension is inhibited; thereby, its release to the atmosphere would prevent its accumulation (Visser et al. 1997; Colmer 2003). Visser and Pierik (2007) proved this idea by assessing 14 species including wetland and non-wetland plants to estimate the internal ethylene concentration in roots. These authors found that wetland species have lower internal concentrations than their relatives from non-flooded sites despite the high variability among species and concluded that the presence of an effective venting system (i.e., aerenchyma) plays a crucial role in avoiding root growth cessation by ethylene accumulation (Visser and Pierik 2007). Ethylene can go upward through the aerenchyma and reach the shoot tissues where it also triggers specific responses like leaf epinasty in sensitive species (Jackson and Campbell 1976), leaf hyponastic growth in tolerant species (upward leaf movement, Polko et al. 2013), and lysigenous aerenchyma in leaf sheaths (Parlanti et al. 2011), among the most conspicuous.

Carbon dioxide accumulates in waterlogged soils and sediments, reaching pressures of up to tens of kPa (Ponnamperuma 1984), and it can flow from roots to shoots through the gas-filled spaces provided by the aerenchymatic tissue. The pressures of carbon dioxide in the aerenchyma lacunae can be several times (>10) higher than those in the atmosphere (Constable et al. 1992). In emergent plant species, such carbon dioxide can come from the sediments and/or the respiration of submerged tissues (roots or rhizomes). In *Typha latifolia*, it was shown that when in the leaf, the increased pressure of carbon dioxide in the aerenchyma does not provoke significant changes in stomatal conductance but allows for photosynthesis of adjacent cells to these gas spaces to be like those of cells close to the epidermis and fueled by carbon dioxide from the atmosphere (Constable and Longstreth 1994). Despite the fact that high levels of photosynthesis fueled by carbon dioxide from aerenchyma would last about a few hours every morning, they might contribute to some extent during the rest of the day when carbon dioxide in aerenchyma is lower (Jackson and Armstrong 1999). Other studies in *Phragmites australis* (Brix 1990), *Scirpus lacustris*, and *Cyperus papyrus* (Singer et al. 1994) suggest that photosynthesis from carbon dioxide in aerenchyma plays a minor role in the carbon economy at plant level (Jackson and Armstrong 1999). In floating leaved species, pioneer work by Dacey and Klug (1982) showed for *Nuphar luteum* in experiments that used $^{14}CO_2$ that a great proportion (85%) of carbon dioxide in aerenchyma lacunae (produced in rhizomes or roots or coming from the sediment) may be fixed by photosynthesis during daylight, while the rest is released to the atmosphere.

Tropospheric methane is the second-most important biogenic greenhouse gas, with a 27 times higher global warming potential than CO_2 on a 100-year time horizon. The three main emission pathways in wetlands are diffusion through the water column, gas bubble release (ebullition), and plant-mediated transport. Recently, a review by Vroom et al. (2022) highlights that plant-mediated transport is the main methane emission pathway in vegetated wetlands. The transport mechanisms include diffusion, pressurized flow, and transpiration-driven flow, the later mainly in wetland trees (Barba et al. 2019).

Diffusion of methane within gas-filled spaces in roots is determined by the presence/absence of a barrier to ROL in the outer cortex, the amount of root porosity (i.e., aerenchyma), the concentrations of methane in soils, and the root surface area through which gases permeate. On the one hand, high rates of radial oxygen loss from the roots to the rhizosphere can reduce methane production and increase its oxidation, which could lessen methane diffusion (Colmer 2003). On the other hand, both oxygen release and methane uptake occur at the permeable root surfaces, so the diffusive transport of methane depends on the methane concentration in the soil and the extent of methane oxidation (see Henneberg et al. 2012 for manipulative experiments in *Juncus effusus*). Young fine roots and tips of old roots lacking ROL barrier are more permeable to methane and are entry points for this gas to the plant. Aerial tissues are not so limiting to methane diffusion because they usually have high gas-filled porosity and permeability. Henneberg et al. (2012) stated that stomatal conductance does not affect the methane diffusion, which is usually released to the atmosphere from the base of stems or petioles above the water level (Harden and Chanton 1994; Vroom et al. 2022). The methane diffusion does not usually follow a diel pattern because methane is mostly independent of stomatal aperture.

Pressurized flows also contribute to methane emissions from plants. Contrary to diffusion, there is a marked diel cycle in methane emissions in plants with pressurized flow (van den Berg et al. 2016). In this case, pressurization of gases (oxygen) within plant tissues is higher when irradiance and temperature are high and humidity is low, so that the maximum venting of gases is attained during the afternoon and the minimum during the night (Brix et al. 1992; Vroom et al. 2022). This cycle affects the methane and CO_2 emissions to the atmosphere as methane accumulates in shoots during the night, when gas flow does not occur, and at sunrise it commences to be released when the day pressurization starts (Chanton et al. 1993; Vroom et al. 2022).

Nitrous oxide is a potent greenhouse gas whose emissions are dependent on microbial production and consumption in the subsurface, and it integrates with the physical water-to-air transfer processes linking its belowground accumulation to atmospheric emissions (Firestone and Davidson 1989). Few studies of nitrous oxide emission through plants have been conducted so far, with most of them focusing on paddy rice (Bhadrachalam et al. 1992), *Sagittaria latifolia*, *Schoenoplectus acutus* (Wang and Reid 2020), *Pontederia cordata*, *Juncus effusus* (Reddy et al. 1989), and woody species from flood-prone areas (Rusch and Rennenberg 1998). The flux of N gases from the soil is greater when rice plants are considered in the system than when there is no crop, which suggests that the nitrous oxide produced in the waterlogged soil is transported through rice plants (Bhadrachalam et al. 1992). A recent study evaluated the significance of nitrous oxide transfer into root gas-filled aerenchyma using wetland mesocosms of *Sagittaria latifolia* and *Schoenoplectus acutus* and found that root-mediated gas transport can account for up to 37% of nitrous oxide removal from the wetland soils (Wang and Reid 2020). The rates of microbial reduction of nitrous oxide varies depending on the soil organic matter and regulates the proportion of nitrous oxide that is transferred to roots. In the case of woody species, it was shown that 3-year-old saplings of *Alnus glutinosa* can emit nitrous oxide through the bark of the stem into the atmosphere when the root zone has concentrations of nitrous oxide higher than those in the atmosphere (Rusch and Rennenberg 1998). Immediately after soil waterlogging, nitrous oxide is emitted from the seedlings' bark at a rate of 350 $\mu mol\ m^{-2}\ h^{-1}$ to later decline to below detection level (<2 $\mu mol\ m^{-2}\ h^{-1}$) after 40 days of water excess. The efflux decreases with increasing stem height and correlates with gas mixing ratios in the soil (nitrous oxide and methane) suggesting diffusion through the aerenchyma as the main mechanism. Therefore, woody species with aerenchyma can serve as conduits for soil-derived nitrous oxide to the atmosphere (Rusch and Rennenberg 1998).

9.5 Future Perspectives

Advances made during the last years about gas transport and plant venting have been enlightening, now illustrated as never before with the use of currently available techniques like noninvasive live imaging systems (e.g., positron-emitting tracer imaging system, Yin et al. 2021). Nevertheless, there are some related topics that have been less explored and could help provide a better understanding of plant critical processes under soil oxygen deficiency, like nutrient acquisition. To illustrate the point, some modelling has been done to explore the potential trade-off between internal aeration capacity and nutrient uptake in rice roots (Kirk 2003) and concluded that for the typical root architecture of current genotypes (aerenchymatous coarse primary root with a tight barrier to ROL conducting oxygen toward short fine, and gas-permeable laterals), there is adequate nitrification of ammonium to nitrate in the rhizosphere to allow half the plant's N to be absorbed (despite denitrification) and to oxidize potential toxins like ferrous iron. Nevertheless, for other species, root architectures, and nutrients, information is scant. Also, the interaction among specialists working at different levels of organization (molecular biologists, plant physiologists, and plant ecologists) should be encouraged by integrating multidisciplinary groups to deal with the increasing demand for flooding-tolerant crops worldwide.

The basic measurements conditions and resulting flow rates from those pressure gradients are also informed.

Acknowledgments I would like to thank Tim Colmer (Univ. Western Australia), Ole Pedersen (Univ. Copenhagen), Lukasz Kotula (Univ. Western Australia), Federico Mollard, and Agustín Grimoldi (Univ. Buenos Aires) for the encouraging discussions about low-oxygen plants' responses over the years. I especially thank Ole for the feedback while preparing this text, Lukasz and Tim for generously sharing the cross sections used for illustrating root aerenchyma, and Lucía Pérez Pizá for improving the artwork of the figures.

References

Abiko T, Kotula L, Shiono K, Malik AI, Colmer TD, Nakazono M (2012) Enhanced formation of aerenchyma and induction of a barrier to radial oxygen loss in adventitious roots of *Zea nicaraguensis* contribute to its waterlogging tolerance as compared with maize (*Zea mays* ssp. mays). Plant Cell Environ 35:1618–1630

Armstrong W (1979) Aeration in higher plants. In: Woolhouse HWW (ed) Advances in botanical research, vol 7. Academic, London, pp 225–332

Armstrong J, Armstrong W (1990) Pathways and mechanisms of oxygen transport in *Phragmites australis*. In: The use of constructed wetlands in water pollution control. Pergamon Press, Oxford, pp 529–533

Armstrong J, Armstrong W (1991) A convective through-flow of gases in Phragmites australis (Cav.) Trin. ex Steud. Aquat Bot 39:75–88

Armstrong J, Armstrong W (1994) A physical model involving Nuclepore membranes to investigate the mechanism of humidity-induced convection in Phragmites australis. Proc R Soc Edinb, Sect B, Biol Sci 102:529–539

Armstrong W, Armstrong J (2014) Plant internal oxygen transport (diffusion and convection) and measuring and modelling oxygen gradients. In: Mancuso, Shabala (eds) Low-oxygen stress in plants. Springer, Vienna, pp 267–297

Armstrong W, Beckett PM (2011) The respiratory down-regulation debate. New Phytol 190:276–278

Armstrong W, Healy MT, Webb T (1982) Oxygen diffusion in pea: I. Pore space resistance in the primary root. New Phytol 91:647–659

Armstrong J, Armstrong W, Beckett PM (1992) *Phragmites australis*: Venturi-and humidity-induced pressure flows enhance rhizome aeration and rhizosphere oxidation. New Phytol 120:197–207

Armstrong W, Brändle R, Jackson MB (1994) Mechanisms of flood tolerance in plants. Acta Bot Neerl 43:307–358

Armstrong W, Armstrong J, Beckett PM (1996a) Pressurised aeration in wetland macrophytes: some theoretical aspects of humidity-induced convection and thermal transpiration. Folia Geobot 31:25–36

Armstrong J, Armstrong W, Beckett PM, Halder JE, Lythe S, Holt R, Sinclair A (1996b) Pathways of aeration and the mechanisms and beneficial effects of humidity-and Venturi-induced convections in *Phragmites australis* (Cav.) Trin. Ex Steud. Aquat Bot 54:177–197

Armstrong J, Lemos EEP, Zobayed SMA, Justin SHFW, Armstrong W (1997) A humidity-induced convective throughflow ventilation system benefits *Annona squamosa* L. explants and coconut calloid. Ann Bot 79:31–40

Armstrong W, Webb T, Darwent M, Beckett PM (2009) Measuring and interpreting respiratory critical oxygen pressures in roots. Ann Bot 103:281–293

Barba J, Bradford MA, Brewer PE, Bruhn D, Covey K, van Haren J, Megonigal JP, Nørgaard Mikkelsen T, Pangala SR, Pihlatie M, Poulter B, Rivas-Ubach A, Schadt CW, Terazawa K, Warner DL, Zhang Z, Vargas R (2019) Methane emissions from tree stems: a new frontier in the global carbon cycle. New Phytol 222:18–28

Beckett PM, Armstrong W, Justin SHFW, Armstrong J (1988) On the relative importance of convective and diffusive gas-flows in plant aeration. New Phytol 110:463–468

Bendix M, Tornbjerg T, Brix H (1994) Internal gas transport in *Typha latifolia* L. and *Typha angustifolia* L. 1. Humidity-induced pressurization and convective throughflow. Aquat Bot 49:75–89

Bhadrachalam A, Chakravorti SP, Banerjee NK, Mohanty SK, Mosier AR (1992) Denitrification in intermittently flooded rice fields and N-gas transport through rice plants. Ecol Bull 42:183–187

Brix H (1990) Gas exchange through the soil-atmosphere interphase and through dead culms of *Phragmites australis* in a constructed reed bed receiving domestic sewage. Water Res 24:259–266

Brix H, Sorrell BK, Orr PT (1992) Internal pressurization and convective gas flow in some emergent freshwater macrophytes. Limnol Oceanogr 37:1420–1433

Brix H, Sorrell BK, Schierup HH (1996) Gas fluxes achieved by in situ convective flow in *Phragmites australis*. Aquat Bot 54:151–163

Chanton JP, Whiting GJ, Happell JD, Gerard G (1993) Contrasting rates and diurnal patterns of methane emission from emergent aquatic macrophytes. Aquat Bot 46:111–128

Colmer TD (2003) Long-distance transport of gases in plants: a perspective on internal aeration and radial oxygen loss from roots. Plant Cell Environ 26:17–36

Colmer TD, Greenway H (2011) Ion transport in seminal and adventitious roots of cereals during O_2 deficiency. J Exp Bot 62:39–57

Constable JV, Longstreth DJ (1994) Aerenchyma carbon dioxide can be assimilated in *Typha latifolia* L. leaves. Plant Physiol 106:1065–1072

Constable JY, Grace JB, Longstreth DJ (1992) High carbon dioxide concentrations in aerenchyma of *Typha latifolia*. Am J Bot 79:415–418

Dacey JW, Klug MJ (1982) Ventilation by floating leaves in *Nuphar*. Am J Bot 69:999–1003

Dedes D, Woermann D (1996) Convective gas flow in plant aeration and thermo-osmosis: a model experiment. Aquat Bot 54:111–120

Firestone MK, Davidson EA (1989) Microbiological basis of NO and N_2O production and consumption in soil. In: Exchange of trace gases between terrestrial ecosystems and the atmosphere, vol 47. Wiley, pp 7–21

Gaynard TJ (1979) Some aspects of internal aeration in wetland plants. (Doctoral dissertation). University of Hull

Gibberd MR, Gray JD, Cocks PS, Colmer TD (2001) Waterlogging tolerance among a diverse range of *Trifolium* accessions is related to root porosity, lateral root formation and 'aerotropic rooting'. Ann Bot 88:579–589

Greenwood DJ (1967) Studies on the transport of oxygen through the stems and roots of vegetable seedlings. New Phytol 66:337–347

Grosse W (1996) Pressurised ventilation in floating-leaved aquatic macrophytes. Aquat Bot 54:137–150

Grosse W, Bauch C (1991) Gas transfer in floating-leaved plants. Vegetatio 97:185–192

Harden HS, Chanton JP (1994) Locus of methane release and mass-dependent fractionation from two wetland macrophytes. Limnol Oceanogr 39:148–154

Henneberg A, Sorrell BK, Brix H (2012) Internal methane transport through *Juncus effusus*: experimental manipulation of morphological barriers to test above-and below-ground diffusion limitation. New Phytol 196:799–806

Jackson MB, Armstrong W (1999) Formation of aerenchyma and the processes of plant ventilation in relation to soil flooding and submergence. Plant Biol 1:274–287

Jackson MB, Campbell DJ (1976) Waterlogging and petiole epinasty in tomato: the role of ethylene and low oxygen. New Phytol 76:21–29

Jensen CR, Stolzy LH, Letey J (1967) Tracer studies of oxygen diffusion through roots of barley, corn, and rice. Soil Sci 103:23–29

Justin SHFW, Armstrong W (1987) The anatomical characteristics of roots and plant response to soil flooding. New Phytol 106:465–495

Kirk GJ (2003) Rice root properties for internal aeration and efficient nutrient acquisition in submerged soil. New Phytol 159:185–194

Konnerup D, Sorrell BK, Brix H (2011) Do tropical wetland plants possess convective gas flow mechanisms? New Phytol 190:379–386

Kotula L, Ranathunge K, Schreiber L, Steudle E (2009) Functional and chemical comparison of apoplastic barriers to radial oxygen loss in roots of rice (*Oryza sativa* L.) grown in aerated or deoxygenated solution. J Exp Bot 60:2155–2167

Kotula L, Clode PL, Striker GG, Pedersen O, Läuchli A, Shabala S, Colmer TD (2015) Oxygen deficiency and salinity affect cell-specific ion concentrations in adventitious roots of barley (*Hordeum vulgare*). New Phytol 208:1114–1125

Leuning R (1983) Transport of gases into leaves. Plant Cell Environ 6:181–194

Manzur ME, Grimoldi AA, Insausti P, Striker GG (2014) Radial oxygen loss and physical barriers in relation to root tissue age in species with different types of aerenchyma. Funct Plant Biol 42:9–17

Mevi-Schutz J, Grosse W (1988) A two-way gas transport system in *Nelumbo nucifera*. Plant Cell Environ 11:27–34

Parlanti S, Kudahettige NP, Lombardi L, Mensuali-Sodi A, Alpi A, Perata P, Pucciariello C (2011) Distinct mechanisms for aerenchyma formation in leaf sheaths of rice genotypes displaying a quiescence or escape strategy for flooding tolerance. Ann Bot 107:1335–1343

Pedersen O, Rich SM, Colmer TD (2009) Surviving floods: leaf gas films improve O_2 and CO_2 exchange, root aeration, and growth of completely submerged rice. Plant J 58:147–156

Pedersen O, Sauter M, Colmer TD, Nakazono M (2021) Regulation of root adaptive anatomical and morphological traits during low soil oxygen. New Phytol 229:42–49

Polko JK, Pierik R, van Zanten M, Tarkowská D, Strnad M, Voesenek LA, Peeters AJ (2013) Ethylene promotes hyponastic growth through interaction with ROTUNDIFOLIA3/CYP90C1 in *Arabidopsis*. J Exp Bot 64:613–624

Ponnamperuma FN (1984) Effects of flooding on soils, vol 10. Academic Press, New York, pp 9–45

Reddy KR, Patrick WH Jr, Lindau CW (1989) Nitrification–denitrification at the plant root–sediment interface in wetlands. Limnol Oceanogr 34:1004–1013

Rusch H, Rennenberg H (1998) Black alder (*Alnus glutinosa* (L.) Gaertn.) trees mediate methane and nitrous oxide emission from the soil to the atmosphere. Plant Soil 201:1–7

Sasidharan R, Voesenek LA (2015) Ethylene-mediated acclimations to flooding stress. Plant Physiol 169:3–12

Shiono K, Yamauchi T, Yamazaki S, Mohanty B, Malik AI, Nagamura Y, Nishizawa NK, Tsutsumi N, Colmer TD, Nakazono M (2014) Microarray analysis of laser-microdissected tissues indicates the biosynthesis of suberin in the outer part of roots during formation of a barrier to radial oxygen loss in rice (*Oryza sativa*). J Exp Bot 65:4795–4806

Singer A, Eshel A, Agami M, Beer S (1994) The contribution of aerenchymal CO_2 to the photosynthesis of emergent and submerged culms of *Scirpus lacustris* and *Cyperus papyrus*. Aquat Bot 49:107–116

Smirnoff N, Crawford RMM (1983) Variation in the structure and response to flooding of root aerenchyma in some wetland plants. Ann Bot 51:237–249

Sorrell BK, Brix H, Orr PT (1997) *Eleocharis sphacelata*: internal gas transport pathways and modelling of aeration by pressurized flow and diffusion. New Phytol 136:433–442

Steinberg SL (1996) Mass and energy exchange between the atmosphere and leaf influence gas pressurization in aquatic plants. New Phytol 134:587–599

Striker GG (2023) An overview of oxygen transport in plants: diffusion and convection. Plant Biol 25:842–847

Striker GG, Insausti P, Grimoldi AA, Vega AS (2007) Trade-off between root porosity and mechanical strength in species with different types of aerenchyma. Plant Cell Environ 30:580–589
van den Berg M, Ingwersen J, Lamers M, Streck T (2016) The role of *Phragmites* in the CH_4 and CO_2 fluxes in a minerotrophic peatland in Southwest Germany. Biogeosciences 13:6107–6119
Visser EJ, Pierik R (2007) Inhibition of root elongation by ethylene in wetland and non-wetland plant species and the impact of longitudinal ventilation. Plant Cell Environ 30:31–38
Visser EJW, Nabben RHM, Blom CWPM, Voesenek LACJ (1997) Elongation by primary lateral roots and adventitious roots during conditions of hypoxia and high ethylene concentrations. Plant Cell Environ 20:647–653
Visser EJW, Colmer TD, Blom CWPM, Voesenek LACJ (2000) Changes in growth, porosity, and radial oxygen loss from adventitious roots of selected mono-and dicotyledonous wetland species with contrasting types of aerenchyma. Plant Cell Environ 23:1237–1245
Vroom RJE, van den Berg M, Pangala SR, van der Schee OE, Sorrell BK (2022) Physiological processes affecting methane transport by wetland vegetation—a review. Aquat Bot 182:103547
Wang S, Reid MC (2020) Kinetics of nitrous oxide mass transfer from porewater into root aerenchyma of wetland plants. J Environ Qual 49:1717–1729
Watanabe K, Takahashi H, Sato S, Nishiuchi S, Omori F, Malik AI, Colmer TD, Mano Y, Nakazono M (2017) A major locus involved in the formation of the radial oxygen loss barrier in adventitious roots of teosinte *Zea nicaraguensis* is located on the short-arm of chromosome 3. Plant Cell Environ 40:304–316
Wegner LH (2010) Oxygen transport in waterlogged plants. In: Waterlogging signalling and tolerance in plants. Springer, Berlin, pp 3–22
Yamauchi T, Nakazono M (2022) Mechanisms of lysigenous aerenchyma formation under abiotic stress. Trends Plant Sci 27:13–15
Yamauchi T, Shimamura S, Nakazono M, Mochizuki T (2013) Aerenchyma formation in crop species: a review. Field Crop Res 152:8–16
Yamauchi T, Tanaka A, Inahashi H, Nishizawa NK, Tsutsumi N, Inukai Y, Nakazono M (2019) Fine control of aerenchyma and lateral root development through AUX/IAA-and ARF-dependent auxin signalling. PNAS 116:20770–20775
Yamauchi T, Pedersen O, Nakazono M, Tsutsumi N (2021) Key root traits of Poaceae for adaptation to soil water gradients. New Phytol 229:3133–3140
Yin YG, Mori Y, Suzui N, Kurita K, Yamaguchi M, Miyoshi Y, Nagao Y, Ashikari M, Nagai K, Kawachi N (2021) Non-invasive imaging of hollow structures and gas movement revealed the gas partial-pressure-gradient-driven long-distance gas movement in the aerenchyma along the leaf blade to submerged organs in rice. New Phytol 232:1974–1984

Part V
Development of Plant Anaerobic Activity

Chapter 10
Anaerobic Germination in Rice

Debabrata Panda, Prafulla K. Behera, and Jijnasa Barik

10.1 Introduction

Rice (*Oryza sativa* L.) is a major staple food crop that has continuously provide global food security. This objective is primarily achieved and fulfilled by adapting high yield rice varieties. Rice of certain varieties has the strength to survive in complete submergence for a period of 1–2 weeks in flash-flood situations. It is also established through research that rice can withstand the vigor of prolonged waterlogged condition even for several months (Singh et al. 2017; Panda and Barik 2021). But unfortunately, most of the modern high-yielding rice varieties are highly susceptible to flooding during germination (Singh et al. 2017). Most of the developed varieties are either completely unable to germinate under anoxic condition or fail to extend the coleoptiles, and they are unable to establish root and shoot under a prolonged oxygen limitation, resulting in crop loss (Magneschi and Perata 2009; Ghosal et al. 2019). Thus, flooding always remained as one of the major constraints to rice production, which has resulted in fluctuations in yield and crop loss mainly in South and Southeast Asian countries, where rice cultivation is normally taken up. A recent study revealed that the frequent flooding events occur due to change in climatic scenarios, and its effect has always become detrimental to rice production. Genetic variation for anaerobic germination exists among different rice genotypes, and in some rice landraces, it is very high. These valuable rice landraces with inherent capacity of anaerobic germination are ideal for studying physiological and molecular mechanism underlying the tolerance.

Though process of transplantation involves tedious labor and requires energy, still the same is commonly adapted in rice cultivation. But in recent past, adaptation of "direct seeding" method has become popular, and in practice, major rice-growing

D. Panda (✉) · P. K. Behera · J. Barik
Department of Biodiversity and Conservation of Natural Resources, Central University of Odisha, Koraput, India

J.-I. Sakagami, M. Nakazono (eds.), *Responses of Plants to Soil Flooding*,
https://doi.org/10.1007/978-981-99-9112-9_10

areas of Southeast Asian countries perceive its various advantages. The direct seeding system of rice is adapted mainly for the reason of early harvest and of high yield with minimum energy input, less labor, less use of water, and mainly on account of operational simplicity (Umarani et al. 2018). But one of the major drawbacks for the wide adaptation of direct seeded rice method is the "poor crop establishment." Therefore, the ability of rice genotypes to survive under anaerobic condition along with subsequent early seedling growth is an essential prerequisite for proper adaptation of direct seeded rice. The tolerance ability of rice genotypes under anoxic condition lowers the risk associated with poor and irregular crop establishment. So the assessment and screening of contrasting rice germplasm for anaerobic germination assume importance and the same need to be addressed which would help in studying various metabolic and signalling pathways associated with the tolerance behavior. More importantly, the improvement of anaerobic germination tolerance ability in high-yielding rice varieties is very much crucial for adaptation of direct seeding system under both lowland rainfed and irrigated ecosystems. Therefore, the present chapter described about an overview of the progress of research achieved on anaerobic germination in rice and development of potential rice genotypes for anaerobic germination tolerance through breeding programs.

10.2 Impact Caused Due to Anoxic Condition During Germination

Rice is a semi-aquatic crop. But it is susceptible to different types of flooding stress, which mainly depends upon the genotype, growth stage, flooding characteristics, and floodwater condition (Mackill et al. 2012). The consequences of flooding usually occur at the first instance during seed germination and at the time of establishment of seedling. Flooding significantly affects the crop establishment and productivity in rice (Miro et al. 2017). This situation is mainly prevalent in rainfed lowland and irrigated areas where flooding occurs soon after direct sowing, leading to anoxic condition while under germination stage (Panda et al. 2017). Under this condition, rice seed is completely inundated with floodwater, resulting in either poor or no germination, and also adversely affects the stand establishment (Magneschi and Perata 2009).

10.3 Approaches Adopted in Rice Under Anaerobic Germination

In case of direct seeding system, anaerobic germination tolerance is considered as key characteristics for rice to successfully germinate (Ismail et al. 2009). Plants have adapted two important survival strategies such as (1) low-oxygen (O_2)

quiescence strategy (LOQS) and (2) low-oxygen (O_2) escape strategy (LOES) in response to submergence stress (Voesenek and Bailey-Serres 2013; Ma et al. 2020). The LOES strategy promotes the mesocotyl and coleoptile elongation during rice germination and early seedling stages, allowing plants to reach the water surface and increase air exchange between aerial and submerged tissues, and the stress-tolerant rice varieties are those that have adopted the LOES strategy in response to submerge stress during germination (Ismail et al. 2009). LOQS, on the other hand, is a strategy that limits the mesocotyl and coleoptile elongation while preserving energy for prolonged submergence stress. Slow coleoptile growth under low-oxygen conditions makes the rice varieties susceptible to submergence stress, which impairs seedling growth development (Fig. 10.1). For instance, cell division is activated during the early stages of germination to increase cell numbers for coleoptiles growth, and subsequent cell expansion contributes to coleoptiles elongation under anaerobic conditions (Takahashi et al. 2011). As a result, coleoptiles elongation is considered as a key criterion for selecting anaerobic germination-tolerant rice varieties (Ma et al. 2020). In addition, rice can sense low O_2 stress and regulate germination and early seedling growth in hypoxic conditions (Magneschi and Perata 2009; Ma et al. 2020). The two adaptive strategies for flooding or submerged tolerance in established rice plants are *SUBMERGENCE 1A* (*SUB1A*)-dependent quiescence and *SNORKEL 1* and *SNORKEL 2* (*SK 1/2*)-dependent escape strategy (Ma et al. 2020).

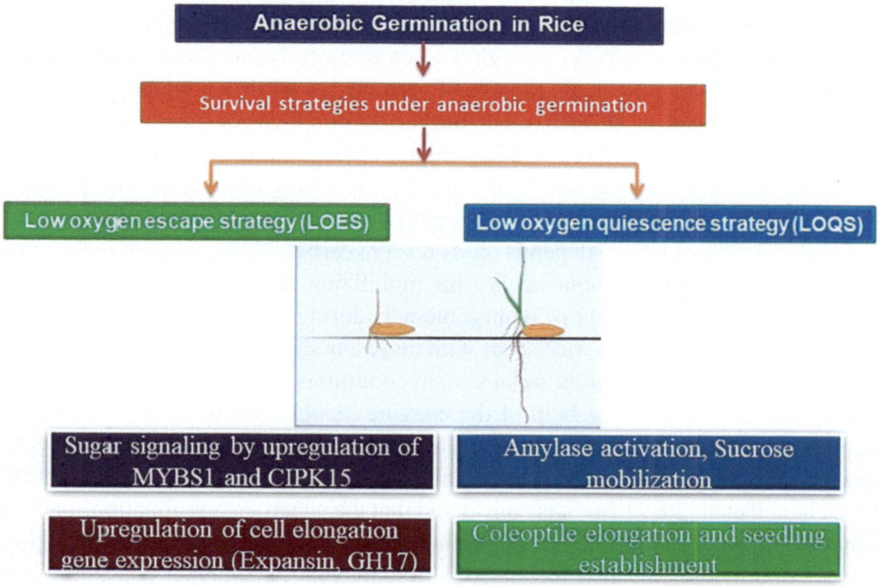

Fig. 10.1 Two of the most important adapted survival strategies of rice seedlings under anaerobic germination such as (1) low-oxygen escape strategy (LOES) and (2) low-oxygen quiescence strategy (LOQS). The tolerant cultivars exhibit phenotypes with faster coleoptiles elongation capacity to reach the water surface for air exchange, while sensitive cultivars adopt the quiescence strategy, which inhibits germination and coleoptiles elongation under low-oxygen conditions

The increased *alcohol dehydrogenase* (*ADH*) gene expression occurs due to *SUB1* gene overexpression for ethanolic fermentation and thereby conferred vegetative stage submergence tolerance in rice by upregulation of gibberellins (GA) signalling regulators such as (a) slender rice-1 (SLR1) and (b) SLR1 like-1 (SLRL1) regulator expression, which further inhibits rice elongation, lowers consumption of carbohydrates, and exhibits strong submergence tolerance in rice (Ma et al. 2020). Rice cultivars with *SUB1* gene could withstand complete submergence for up to 2 weeks (Xu et al. 2006). However, in case of deepwater rice, the SK 1/2-dependent escape strategy enhances internode elongation in responses to flooding stress by stimulating GA biosynthesis, allowing rice to grow upward to water surface for air exchange (Ma et al. 2020).

10.4 Mechanism Associated with Anaerobic Germination Tolerance

The potential capability of some of the rice genotypes to germinate and grow and withstand under oxygen deprived condition is termed as "anaerobic germination tolerance" (Ella and Setter 1999; Illangakoon et al. 2019). Tolerant rice genotypes are able to sustain the low-oxygen condition by developing its own adaptive strategies. The rice seeds under water do not get ample oxygen and unable to generate adequate energy for which threat is caused in its sustenance. Hence, the coleoptiles emerge out of the water to get enough oxygen to meet its metabolic requirements. Rice is one of the major cereals capable of germination while completely inundated under water (Angaji et al. 2010), but further development of plants is limited to coleoptile elongation (Biswas and Yamauchi 1997). Rapid coleoptile elongation facilitates and provides air and light to the germinating seed, and such aeration would help in the development of the embryo. At the initial stages, while under water, the growing embryo depends on its reserve carbohydrates present in the seed endosperm. Rice has a unique ability for mobilizing the starch of endosperm to fulfil the energy requirement of embryonic axis during oxygen deficient condition (Hsu and Tung 2015). Thus, rice seeds with more amount of stored carbohydrate are considered to be more tolerant to anaerobic conditions. The starch mobilization is mainly carried out with the help of the enzyme *alpha-amylase* by degradation of starch into fermentable carbohydrates. The role of *alpha-amylase* in starch breakdown under oxygen limitation was established by different studies (Hwang et al. 1999; Ismail et al. 2012). So, it is considered that the anaerobic germination-tolerant genotypes have better ability of degrading starch into usable sugars. The higher activity of amylase was also correlated with maintenance of higher concentrations of soluble sugars, greater degradation of starch, better growth of root and shoot, and greater survival of seedling under anaerobic condition (Ismail et al. 2009). In addition, the enzymes linked with alcoholic fermentation and glycolysis are upregulated for energy generation during anaerobic condition. Oxygen limitation is involved with increased ethanol fermentation, which is catalyzed by the enzyme *alcohol dehydrogenase* and *pyruvate decarboxylase*. In anoxic condition, increased activity of *lactate dehydrogenase* is noticed, which is associated with lactate fermentation. However, the production of ATP was 18-fold less efficient in anoxic condition when compared to aerobic condition (Ismail et al. 2012). The elongation of coleoptiles and mesocotyls in early growth is promoted by the plant hormone "ethylene," which is beneficial to the germinating seed in anaerobic condition. But the ethylene in case of flash-flood situation is detrimental to plants as it causes lodging due to excessive internode elongation. Studies revealed the importance of ethylene on coleoptiles and shoot elongation during germination under anoxic/hypoxic condition (Ismail et al. 2009). Different group of proteins like "expansions" and "peroxidases" play important role in regulation of cell wall expansion and growth under anaerobic condition in germinating seeds (Ismail et al. 2012). Thus, it is believed that various mechanisms are responsible to withstand the low oxygen stress during germination and early seedling growth. Extensive study is required for identification of appropriate mechanisms as well as identification of genes underlying the tolerance, which could be used for the development of tolerant genotypes in the near future.

10.5 Screening for Anaerobic Germination Tolerance in Rice

The ability to germinate under oxygen-deprived condition varies greatly within different rice cultivars. Two hundred fifty-eight rice accessions from the gene bank of IRRI and four hundred four genotypes from the International Network for the Genetic Evaluation of Rice (INGER) were screened for anaerobic germination, and from the experiment, 12 rice lines were identified as tolerant genotypes with 54–78% range of coleoptiles emergence when compared with susceptible varieties (Yamauchi et al. 1993). Another experiment carried out by Jiang et al. (2004) taking 359 *indica* and *japonica* rice accessions where coleoptiles elongation was the major selection criterion for identification of suitable genotypes. The authors observed significant variation in coleoptiles growth after 5 days of exposure to treatment condition. More than 8000 rice varieties of IRRI Genetic Resources Centre were screened for anaerobic germination at the initial stages, and it was observed that only 0.23% of all rice accessions had 70% of survival. Further evaluation on replicated trials reduced the initial percentage to again 0.06%. Few rice varieties like Khaiyan, Khao Hlan On, Ma-Zhan Red, and Nanhi were identified as anaerobic germination-tolerant genotypes through standardized screening procedure at the International Rice Research Institute (Angaji et al. 2010). Rice genotypes (359 in number) were studied for anaerobic germination under 20 cm of water depth by considering the key trait i.e. 'coleoptiles elongation' (Jiang et al. 2004). Eighty-eight rice landraces were also screened for anaerobic germination potential by taking different morphophysiological traits as selection criteria (Barik et al. 2019), and the results showed three rice landraces had superior anaerobic germination ability and seedling growth among all the studied rice genotypes. Screening and assessment of genetic diversity of 107 rice varieties were studied under oxygen stress for

14 days, and several suitable tolerant rice genotypes were identified (Sudeepthi et al. 2019, 2020; Panda et al. 2021). Hence, more researches are required to be carried out for the assessment and exploration of genetic variability of anaerobic germination from diverse rice germplasms.

10.6 The Metabolic Adaption and Molecular Regulatory Pathways of Anaerobic Germination in Rice

Aerobic respiration is severely inhibited in rice under submerged germination due to lack of water O_2, resulting in differential expression of genes that contribute to activating the essential mechanisms, as a result, shifting the metabolic processes to anaerobic respiration, which is related to energy production and utilization in the germinating seeds (Ma et al. 2020). Fermentation refers to these alternative pathways to aerobic respiration, which include alcoholic fermentation, alanine pathway, and lactic acid fermentation (Sousa and Sodek 2002; Ma et al. 2020). Meanwhile, the oxidative phosphorylation and TCA cycle are inhibits, and ATP production is shifted from electron transport chains to glycolysis and ethanol fermentation (Ma et al. 2020). As a result, anaerobic fermentation produces less net energy than aerobic respiration. Although ethanol fermentation produces only a small amount energy, it is critical for seedling germination and growth under submergence conditions (Ma et al. 2020). The rapid consumption of soluble carbohydrate during germination and seedling growth stages of cereal grains causes sugar starvation and activation of Amy biosynthesis in the scutellum, which is mediated by the CIPK15–SnRK1A–MYBS1 signalling pathways (Yu et al. 1996; Ma et al. 2020). Only rice produces a complete set of enzymes to degrade starch under anaerobic germination conditions. Rice promotes coleoptiles elongation to survive anaerobic stress by having a higher Amy activity for the degradation of starch to produce energy in germinating seeds (Ma et al. 2020).

The major signals for rice responsiveness under submerged/anaerobic germination are oxygen and sugar deficit. O_2 is considered as a critical component for various useful biochemical activities occurring in rice cells (Ma et al. 2020). When rice germinates in responses to submerged situation, it detects hypoxic stress and activates the appropriate regulatory pathways to adapt to the harsh environment (Ma et al. 2020). Calcium (Ca^{2+}) signalling pathway acts as a secondary messenger in response to different abiotic stresses (Das and Pandey 2010). The calcineurin B-like (CBL) proteins represent as Ca^{2+}-binding proteins and Ca^{2+} ion sensors in plants and form the CBL-CIPK complex by interacting with CBL protein-interacting protein kinases (CIPKs; Ser/Thr protein kinases), decoding Ca^{2+} signalling under different abiotic stresses (Albrecht et al. 2001; Luan 2009; Ma et al. 2020). In recent past, the study on rice by Lu et al. (2007) and Lee et al. (2009) reported about the CBL/CIPK15–SnRK1A–MYBS1-mediated sugar-signalling cascade that is important in low-O_2 tolerance during rice seed germination. CIPK15 is required for the expression of genes encoding Amy and ADH, and CBL4 has been shown to be induced by low oxygen levels, and to interact strongly with the CIPK15 protein, further when CBL4 is silenced, the expression of Amy3 is reduced (Ma et al. 2020). These findings suggest that the CBL4/CIPK15 complex is important in the sugar starvation signalling pathway and may help rice germinate under hypoxia conditions (Ma et al. 2020). The OsCBL10 promoter, a member of CBL family, is linked with anaerobic germination, and seedling growth tolerance has been reported by Ye et al. (2018). SnRK1A, SNF1, and AMPK are metabolic sensors that monitor cellular carbohydrate status or ATP level, which regulates carbon and nitrogen metabolism, and are required for growth and development (Ma et al. 2020). SnRK1A functions as a conserved energy and stress sensor under submergence conditions and is involved in sugar starvation and O_2 deficiency signalling pathways downstream of the CBL/CIPK15 complex (Lee et al. 2009). Furthermore, SnRK1A-interacting negative regulators interacting with SnRK1A negatively regulate and modulate nutrient starvation signalling and its downstream pathways, compromising anaerobic germination and seedling growth in rice under submergence stress (Ma et al. 2020). In addition, sugar deprivation/low O_2 signals caused by submergence activate *CIPK15*, which then activates SnRK1A protein accumulation. In response to submergence stress, SnRK1A accumulation initiates a SnRK1A-dependent signalling cascade. The downstream sugar signalling cascade of SnRK1A-MYBS1-Amy 3 promotes starch degradation and sugar mobilization, indicating that the CIPK15-SnRK1A-MYBS1-mediated sugar sensing pathway is important for rice seed germination under submerged condition (Ye et al. 2018; Ma et al. 2020).

10.7 QTLs Associated with Anaerobic Germination

In the past three decades, breeding for enhanced anaerobic germination and other essential traits has been tried by several researchers (Yamauchi et al. 1993, 1994; Redona and Mackill 1996; Yamauchi and Winn 1996; Biswas 1997), but the progress achieved was very poor because of inadequate information on genetics of anaerobic germination tolerance, complexity of its genetic backgrounds, lack of effective screening procedures, etc. In one study, ‘Khao Hlan On’ (a potential anaerobic tolerant donor) was crossed with ‘IR 64’, a widely cultivated mega-rice variety, as a recurrent parent, and the resulting BC_2F_2 population was further evaluated for anaerobic germination potential (Angaji et al. 2010) (Table 10.1). In this experiment, five different quantitative trait locus (QTLs) were identified, one each on chromosome number 1 (*qAG1.2*), chromosome number 3 (*qAG3.1*), and chromosome number 7 (*qAG7.2*) and two QTLs on chromosome number 9 (*qAG9.1*) and (*qAG9.2*), explaining the phenotypic variation from 17.9 to 33.5%. Among the five QTLs, the largest QTL (*qAG9.2*) was present on the long arm of chromosome number 9, with the highest phenotypic variation and LOD score of 33.5% and 20.3%, respectively. Through further studies, cloning and functional validation of the gene

Table 10.1 QTLs associated with anaerobic germination in rice

Sl no.	QTL	Marker interval	Chromosome	References
1	*qAG7.1*	RM3583–RM21427	7	Septiningsih et al. (2013)
2	*qAG7.2*	RM7338–RM346	7	
3	*qAG7.3*	RM21803–RM234	7	
4	*qAG2*	RM263–RM5378	2	
5	*qAG5*	RM5361	5	
6	*qAG6*	RM204–RM402	6	
7	*qAG9*	RM553–RM3808	9	
8	*qAG12*	RM313–RM28766	12	
9	*qAG1*	RM312	1	
10	*qAG12*	RM28759	12	
11	*qAG7*	wd7000465–id7002784	7	Angaji (2008)
12	*qAG11*	id11009201–id11010245	11	
13	*qAG9.2*	RM3769–RM105	9	Angaji et al. (2010)
14	*qAG9.1*	RM8303–RM5526	9	
15	*qAG7.2*	RM21868–RM172	7	
16	*qAG1.2*	RM11125–RM104	1	
17	*qAG1.1*	RM582–RM10713	1	
18	*qAG2.1*	RM327–RM6318	2	
19	*qAG3.1*	RM7097–RM520	3	
20	*qAG7.1*	RID12[i]–RM5606	7	
21	*qAG8.1*	RM210–RM149	8	
22	*qAG5*	RM405–RM169[a]–RM249	5	Jiang et al. (2006)
23	*qAG11*	RM21–RM254[a]–RM224	11	
24	*qAG3*	Id3002377–id3004190	3	Baltazar et al. (2019)
25	*qAG7.1*	Id7000519–id7002260	7	
26	*qAG7.2*	Id7002427–id7003359	7	
27	*qAG7.3*	Id7003853–id7004429	7	

encoding a "trehalose-6-phosphate phosphatase" (*OsTPP*) underlying the quantitative trait locus (*qAG9.2*) has been carried out (Kretzschmar et al. 2015). The *OsTPP* gene is linked with "trehalose-6-phosphate" (T6P) metabolism, helping in mobilization of starch and faster coleoptiles elongation under anoxic state, leading to enhanced tolerance to anaerobic germination (Singh et al. 2017; Loreti et al. 2018). Mapping of QTL *qAG9.2* has also been reported in several studies (Jiang et al. 2004, 2006; Septiningsih et al. 2013; Baltazar et al. 2014). Jiang et al. (2006) also reported two QTLs on chromosome number 5 and chromosome number 11 having phenotypic variance of 11–15.5% (Table 10.1). Baltazar et al. (2019) observed four QTLs relating to anaerobic germination, i.e., *qAG7.1*, *qAG7.2*, and *qAG7.3* located on chromosome 7 and *qAG3* located on chromosome 3 from IR64/Kharsu 80A $F_{2:3}$ mapping population (Table 10.1). Several studies were carried out to know the genetics of anaerobic germination tolerance by different workers around the world (Baltazar et al. 2014; Hsu and Tung 2015; Zhang et al. 2017; Ghosal et al. 2019;

Yang et al. 2019; Jeong et al. 2020; Nishimura et al. 2020). Some of the important QTLs related to anaerobic germination in rice reported by several researchers have been presented in Table 10.1.

10.8 Breeding for Anaerobic Germination in Rice

Rice is able to withstand the anaerobic condition up to some extent, which is limited to elongation of coleoptiles and partial growth. Therefore, breeding new varieties with higher anaerobic germination potential will help avoid many problems currently associated with direct seeded rice system. Breeding for increased seedling vigor and enhanced anaerobic germination potential has been investigated in rice earlier by various researchers (Yamauchi et al. 1993, 1994; Biswas and Yamauchi 1997; Angaji et al. 2010), but the progress achieved was slow due to lack of superior anaerobic germination tolerant donors and limited information on genetics of tolerance ability (Jiang et al. 2004). Genes for anaerobic germination tolerance have been identified and utilized for development of rice variety with enhanced anaerobic germination potential (Baltazar et al. 2014). Molecular breeding for tolerance to anaerobic germination in rice was started using *qAG9–2* (*AG1*) identified from 'Khao Hlan On' (Kretzschmar et al. 2015). This quantitative trait locus was then incorporated into some improved varieties or their *Sub1* derivatives like IR64, IR64-*Sub1*, PSB Rc18-*Sub1*, and PSB Rc82 (Septiningsih et al. 2013). The traditional rice cultivar 'Ma-Zhan Red' was used as a donor in introgression of *qAG7–1* (*AG2*) locus to popular high-yielding rice varieties in several studies. However, identification of new rice genotypes with increased tolerance to flooding during germination would help the current breeding programs for anaerobic germination.

10.9 Conclusion

The progress achieved through several researches since decades on anaerobic germination in rice has enhanced our knowledge regarding the subject. Still, there exist many areas that need to be addressed for better crop management to increase food production. Hence, research should focus more on study of "anaerobic germination" to unravel the detailed mechanisms concerning the survival of rice seed under low-oxygen condition. The world is facing climatic variation due to several contingencies, causing severe fluctuation in weather pattern. This has taken rice growers to face force majeure conditions of unprecedented rains and floods, etc. So it is highly essential to search for new sources of genes that can offer greater tolerance under changing climatic conditions. The identified QTLs to anaerobic germination should be properly investigated for their utilization in molecular breeding programs. In addition to that, rice seeds with superior anaerobic germination potentiality are highly essential for direct seeded rice system in irrigated as well as rainfed and flood-prone areas. Thus, breeding of rice cultivars with high level of anaerobic germination is crucial for successful application of direct seeded system.

References

Albrecht V, Ritz O, Linder S, Harter K, Kudla J (2001) The NAF domain defines a novel protein–protein interaction module conserved in Ca^{2+}-regulated kinases. EMBO J 20:1051–1063

Angaji SA (2008) Mapping QTLs for submergence tolerance during germination in rice. Afr J Biotechnol 7(15):2551–2558

Angaji SA, Septiningsih EM, Mackill DJ, Ismail AM (2010) QTLs associated with tolerance of flooding during germination in rice (*Oryza sativa* L.). Euphytica 172:159–168

Baltazar MD, Ignacio JCI, Thomson MJ, Ismail AM, Mendioro MS, Septiningsih EM (2014) QTL mapping for tolerance of anaerobic germination from IR64 and the aus landrace Nanhi using SNP genotyping. Euphytica 197:251–260

Baltazar MD, Ignacio JCI, Thomson MJ, Ismail AM, Mendioro MS, Septiningsih EM (2019) QTL mapping for tolerance to anaerobic germination in rice from IR64 and the aus landrace Kharsu 80A. Breed Sci 69(2):227–233

Barik J, Kumar V, Lenka SK, Panda D (2019) Genetic potentiality of lowland indigenous *indica* rice (*Oryza sativa* L.) landraces to anaerobic germination potential. Plant Physiol Rep 24(2):249–261

Biswas JK, Yamauchi M (1997) Mechanism of seedling establishment of direct-seeded rice (*Oryza sativa* L.) under lowland conditions. Bot Bull Acad Sin 38:29–32

Das R, Pandey GK (2010) Expressional analysis and role of calcium regulated kinases in abiotic stress signaling. Curr Genomics 11:2–13

Ella ES, Setter TL (1999) Importance of seed carbohydrates in rice seedling establishment under anoxia. VI Symposium on stand establishment and ISHS seed symposium 504

Ghosal S, Casal C, Quilloy FA, Septiningsih EM, Mendioro MS, Dixit S (2019) Deciphering genetics underlying stable anaerobic germination in rice: phenotyping, QTL identification, and interaction analysis. Rice 12(1):50

Hsu SK, Tung CW (2015) Genetic mapping of anaerobic germination-associated QTLs controlling coleoptile elongation in rice. Rice 8(1):1–12

Hwang SY, Thomas BR, Rodriguez RL (1999) Differential expression of rice alpha-amylase genes during seedling development under anoxia. Plant Mol Biol 40(6):911–920

Illangakoon TK, Marambe B, Keerthisena RSK, Kumar V (2019) Importance of anaerobic seed germination and seedling development in direct-seeded rice with special reference to Sri Lanka. Sri Lanka J Food Agric 5(2):47–57

Ismail AM, Ella ES, Vergara GV, Mackill DJ (2009) Mechanisms associated with tolerance to flooding during germination and early seedling growth in rice (*Oryza sativa*). Ann Bot 103(2):197–209

Ismail AM, Johnson DE, Ella ES, Vergara GV, Baltazar AM (2012) Adaptation to flooding during emergence and seedling growth in rice and weeds, and implications for crop establishment. AoB Plants 2012:pls019

Jeong JM, Cho YC, Jeong JU, Mo YJ, Kim CS, Kim WJ, Baek MK, Kim SM (2020) QTL mapping and effect confirmation for anaerobic germination tolerance derived from the *japonica* weedy rice landrace PBR. Plant Breed 139(1):83–92

Jiang L, Hou M, Wang C, Wan J (2004) Quantitative trait loci and epistatic analysis of seed anoxia germinability in rice (*Oryza sativa*). Rice Sci 11(5–6):238–244

Jiang L, Liu S, Hou M, Tang J, Chen L, Zhai H, Wan J (2006) Analysis of QTLs for seed low temperature germinability and anoxia germinability in rice (*Oryza sativa* L.). Field Crop Res 98(1):68–75

Kretzschmar T, Pelayo MAF, Trijatmiko KR, Gabunada LFM, Alam R, Jimenez R, Mendioro MS, Slamet-Loedin IH, Sreenivasulu N, Bailey-Serres J, Ismail AM, Mackill DJ, Septiningsih EM (2015) A trehalose-6-phosphate phosphatase enhances anaerobic germination tolerance in rice. Nat Plants 1:15124

Lee KW, Chen PW, Lu CA, Chen S, Ho TH, Yu SM (2009) Coordinated responses to oxygen and sugar deficiency allow rice seedlings to tolerate flooding. Sci Signal 2:ra61

Loreti E, Valeri MC, Novi G, Perata P (2018) Gene regulation and survival under hypoxia requires starch availability and metabolism. Plant Physiol 176(2):1286–1298

Lu CA, Lin CC, Lee KW, Chen JL, Huang LF, Ho SL, Liu HJ, Hsing YI, Yu SM (2007) The SnRK1A protein kinase plays a key role in sugar signaling during germination and seedling growth of rice. Plant Cell 19:2484–2499

Luan S (2009) The CBL-CIPK network in plant calcium signaling. Trends Plant Sci 14:37–42

Ma M, Cen W, Li R, Wang S, Luo J (2020) The molecular regulatory pathways and metabolic adaptation in the seed germination and early seedling growth of rice in response to low O_2 stress. Plant Theory 9:1363

Mackill DJ, Ismail AM, Singh US, Labios RV, Paris TR (2012) Development and rapid adoption of submergence-tolerant (Sub1) rice varieties. Adv Agron 115:299–352

Magneschi L, Perata P (2009) Rice germination and seedling growth in the absence of oxygen. Ann Bot 103(2):181–196

Miro B, Longkumer T, Entila FD, Kohli A, Ismail AM (2017) Rice seed germination underwater: morpho-physiological responses and the bases of differential expression of alcoholic fermentation enzymes. Front Plant Sci 8:1857

Nishimura T, Sasaki K, Yamaguchi T, Takahashi H, Yamagishi J, Kato Y (2020) Detection and characterization of quantitative trait loci for coleoptile elongation under anaerobic conditions in rice. Plant Prod Sci 23(3):374–383

Panda D, Barik J (2021) Flooding tolerance in rice: focus on mechanisms and approaches. Rice Sci 28(1):43–57

Panda D, Rao DN, Das KK, Sarkar RK (2017) Role of starch hydrolytic enzymes and phosphatases in relation to under water seedling establishment in rice. Indian J Plant Physiol 22:279–286

Panda D, Barik J, Sarkar RK (2021) Recent advances of genetic resources, genes and genetic approaches for flooding tolerance in rice. Curr Genomics 22:41–58

Redona E, Mackill D (1996) Genetic variation for seedling vigor traits in rice. Crop Sci 36(2):285–290

Septiningsih EM, Ignacio JCI, Sendon PMD, Sanchez DL, Ismail AM, Mackill DJ (2013) QTL mapping and confirmation for tolerance of anaerobic conditions during germination derived from the rice landrace Ma-Zhan red. Theor Appl Genet 126(5):1357–1366

Singh A, Septiningsih EM, Balyan HS, Singh NK, Rai V (2017) Genetics, physiological mechanisms and breeding of flood-tolerant rice (*Oryza sativa* L.). Plant Cell Physiol 58(2):185–197

Sousa CAFD, Sodek L (2002) The metabolic response of plants to oxygen deficiency. Braz J Plant Physiol 14:83–94

Sudeepthi K, Srinivas T, Kumar BNVSRR, Jyothula DPB, Nafeez Umar SK (2019) Screening for tolerance to anaerobic germination in rice (*Oryza sativa* L.). Int J Curr Microbiol Appl Sci 8(12):2527–2538

Sudeepthi K, Srinivas T, Kumar BR, Jyothula D, Umar SN (2020) Genetic divergence studies for anaerobic germination traits in rice (*Oryza sativa* L.). Curr J Appl Sci Technol 39(1):71–78

Takahashi H, Saika H, Matsumura H, Nagamura Y, Tsutsumi N, Nishizawa NK, Nakazono M (2011) Cell division and cell elongation in the coleoptile of rice alcohol dehydrogenase 1-deficient mutant are reduced under complete submergence. Ann Bot 108:253–261

Umarani E, Hemalatha V, Bhadana VP, Senguttuvel P, Subbarao LV, Neeraja CN, Yugandhar RP, Kota S (2018) Screening of rice (*Oryza sativa* L.) genotypes for anaerobic germination. Int J Pure Appl Biosci 6(5):1318–1325

Voesenek LACJ, Bailey-Serres J (2013) Flooding tolerance: O_2 sensing and survival strategies. Curr Opin Plant Biol 16:647–653

Xu K, Xu X, Fukao T, Canlas P, Maghirang-Rodriguez R, Heuer S, Ismail AM, Bailey-Serres J, Ronald PC, Mackill DJ (2006) Sub1A is an ethylene-response-factor-like gene that confers submergence tolerance to rice. Nature 442:705–708

Yamauchi M, Winn T (1996) Rice seed vigor and seedling establishment in anaerobic soil. Crop Sci 36(3):680–686

Yamauchi M, Aguilar AM, Vaughan DA, Seshu DV (1993) Rice (*Oryza sativa* L.) germplasm suitable for direct sowing under flooded soil surface. Euphytica 67:177–184

Yamauchi M, Herradura PS, Aguilar AM (1994) Genotype difference in rice post-germination growth under hypoxia. Plant Sci 100(1):105–113

Yang J, Sun K, Li D, Luo L, Liu Y, Huang M, Yang G, Liu H, Wang H, Chen Z, Guo T (2019) Identification of stable QTLs and candidate genes involved in anaerobic germination tolerance in rice via high-density genetic mapping and RNA-Seq. BMC Genomics 20(1):1–15

Ye NH, Wang FZ, Shi L, Chen MX, Cao YY, Zhu FY, Wu YZ, Xie LJ, Liu TY, Su ZZ, Xiao S, Zhang H, Yang J, Gu HY, Hou XX, Hu QJ, Yi HJ, Zhu CX, Zhang J, Liu YG (2018) Natural variation in the promoter of rice calcineurin B-like protein10 (OsCBL10) affects flooding tolerance during seed germination among rice subspecies. Plant J 94:612–625

Yu SM, Lee YC, Fang SC, Chan MT, Hwa SF, Liu LF (1996) Sugars act as signal molecules and osmotic to regulate the expression of alpha-amylase genes and metabolic activities in germinating cereal grains. Plant Mol Biol 30:1277–1289

Zhang M, Lu Q, Wu W, Niu X, Wang C, Feng Y, Xu Q, Wang S, Yuan X, Yu H, Wang Y, Wei X (2017) Association mapping reveals novel genetic loci contributing to flooding tolerance during germination in *indica* rice. Front Plant Sci 8:678

Chapter 11
Plant Morpho-Physiological Responses to Changes in the Soil Water Status

Phanthasin Kanthavong and Jun-Ichi Sakagami

11.1 The Impact of Climate Change on Rainfed Agriculture and Global Food Security

According to the United Nations (2019), the world's population is expected to grow by more than 2 billion by 2050. The distribution of new growth will be about 50% in sub-Sahara African and 30% in South and Southeast Asia, notably, these regions heavily rely on rainfed agriculture, particularly for cereal crops.

Climate change and weather disasters are major causes of reductions in agricultural productivity, especially rainfed agriculture where the rainfall is the water source for crops. Climate change is expected to drive higher average temperatures, heat waves, droughts, and heavy rain, which are likely to occur more often and for greater duration (Berg et al. 2013; Giorgi et al. 2019). The exact extent of the combination of these factors for any given location remains highly uncertain (IPCC 2014). To address climate change, global farmers must produce approximately 3.0 billion tons of cereal to feed the Earth's population, which is expected to reach nearly 8.0 billion. Ensuring and sustaining a global average cereal yield of 4 tons per hectare is of utmost important for sustainable food security worldwide (Pareek et al. 2009). Despite these challenges, rainfed agriculture will continue to play a dominant role in providing food security and generating livelihoods, particularly in poor countries (Rockström et al. 2010).

Under rainfed agriculture, there are various types of abiotic stresses including drought, flood, salinity, extreme temperatures, chemical toxicity, and oxidative stress that plants encounter between seedling and harvesting stages. Among the

P. Kanthavong (✉)
National Agriculture and Forestry Research Institute (NAFRI), Vientiane, Lao PDR

J.-I. Sakagami
Faculty of Agriculture, Kagoshima University, Kagoshima, Japan

J.-I. Sakagami, M. Nakazono (eds.), *Responses of Plants to Soil Flooding*,
https://doi.org/10.1007/978-981-99-9112-9_11

abiotic stress, drought and flooding are critical challenging issues for crop production (Pareek et al. 2020). When crop plants face water stress, their hormonal responses trigger morphological and physiological modification in both roots and shoots. These phenotypic modifications are the survival or adaptation strategies of crop plants to cope with water stresses. In this chapter, we delve into the effects of changing soil water status, transitioning from aerobic to anaerobic conditions (soil drying-waterlogging scenarios) on the morphological and physiological changes in crop plants. Additionally, we explore the relationship between root and shoot in terms of exchangeable gases.

11.2 The Effect of Changing Soil Water Status on Modification of Root and Shoot

The plant root system is an important part that absorbs water and nutrients from the soil for sustainable crop production (Lynch 1995), particularly in rainfed ecosystems that suffers from the increasing frequency of erratic rainfall patterns due to climate change. The optimization of root architecture for resource capture is vital for enabling the next Green Revolution (Den Herder et al. 2010; Lynch 2007). Thus, scientists have been looking for optimum root development that may help plants use water more efficiently, better withstand drought conditions, and survive under flood conditions (Comas et al. 2013; Suralta et al. 2010; Vadez 2014; Wasson et al. 2012; Weaver et al. 1922; Yamauchi et al. 2018). The growth and development of roots are under genetic control, but the genotypic performance of root distribution is highly influenced by abiotic stress (Fry et al. 2018; Grossman and Rice 2012; McMichael and Quisenberry 1993).

Functional roots supply water-associated nutrients to promote stem elongation, leaf expansion, and consequent shoot biomass and productivity. On the other hand, leaves house the stomata, through which plants uptake CO_2. Simultaneously, leaves release water via transpiration to maintain balanced leaf temperature. Another important leaf function is light absorption which occurs during photosystem (PSII). Under water stresses, therefore, morphological leaf modification could be a survival or adapted strategy.

11.3 Root and Leaf Response to Waterlogging

Waterlogging refers to a situation where the soil becomes submerged due to an increase in groundwater level or excessive rainfall. In this condition, the root-zone of plants is flooded, limiting the amount of oxygen available to plant roots. In the field conditions, waterlogging is caused by heavy precipitation, leading the saturation of soil surface with the water. The severity of stress and response elicited depends upon genotype, growth stage organ, and other factors including duration, light availability, temperature, humility, and the amount of carbohydrate storage in the plant body (Sasidharan et al. 2017).

Excess water overcomes the porous soil causing deficient oxygen (O_2) and carbon dioxide in plant root-zone. Consequenly, flooded roots shift toward anaerobic respiration, and resulting in reduced chemical production necessary for generate energy (ATP) required for cell metabolism (Loreti and Perata 2020). Most plants are sensitive to waterlogging, as it causes low oxygen concentration in root cortex of sensitive species, for example, maize and chickpeas (Colmer et al. 2020; Pedersen et al. 2021). In general, root is the first organ response to change soil moisture status (Khanthavong et al. 2021; Yamauchi et al. 2021). Climate change exacerbates waterlogging, affecting even non-wetland plants. These plants face internal oxygen deficiency due to reduced gas exchange. To survive under waterlogged conditions, plants have evolved the ability to transport oxygen from aerial parts to root tip (Armstrong 1979).

There are many evidences that show plant response to waterlogging by changing root system architecture and root morphology for their survival or adaptive strategies. Most of the changing root system architecture and morphology approaches to increase oxygen use efficiency for root respiration. The rice is a semi-aquatic plant and well adapts to waterlogging (Sakagami and Kawano 2011). Unlike some other crops, rice exhibits less pronounced change in its entire root system architecture and can elongate its root to deeper soil layers. For instance, maize (*Zea may* L.) experiences more significant alteration in root system architecture under waterlogging (Abiko et al. 2012; Yamauchi et al. 2019), sorghum (*bicolor* Moench) (Khanthavong et al. 2021; Promkhambut et al. 2010; Zhang et al. 2019a), and wheat (*Triticum aestivum*) (Yamauchi et al. 2019) also respond differently to waterlogging. After prolonged waterlogging, susceptible species experience severe internal oxygen deficiency, leading to oxidative stress at the root tip (hypoxia). This condition can cause root death at deeper soil layer (Loreti and Perata 2020).

Thereafter, plants activate the hormone, namely ethylene, which functions to regulate adventitious roots (Raskin and Kende 1984; Vidoz et al. 2010). The emergence of adventitious roots is a key response to waterlogging conditions in many plant species (Steffens and Rasmussen 2016), including rice (Lorbiecke and Sauter 1999), tomato (Negi et al. 2010), maize (Mano et al. 2005), and wheat (Nguyen et al. 2018). Ethylene not only promotes the adventitious roots, but it also induces programmed cell death (aerenchyma formation) in mostly adventitious roots associated with radial oxygen loss (ROL) barrier in wetland species (Colmer 2003; Colmer et al. 2006), where it is highly concentrated on adventitious roots. Aerenchyma formation is the principal channel for oxygen diffusion from atmospheric transport to root tip. In rice, aerenchyma formation appears under aerobic conditions. However, in response to waterlogging, aerenchyma formation can be induced within 12 h, resulting in high percentage of porosity (ranging from 30 to 40%), depending on a specific genotype (Armstrong 1979; Colmer 2003; Colmer et al. 2006). This evidence demonstrates why rice can overcome root hypoxia. Other root modifications to mitigate root hypoxia induce changes in root angle, thickness, shorter root distance from the root–shoot junction, and fewer root branches. Due to the critical oxygen concentration existing in deeper soil layers compared to shallow soil layers, newly emerged adventitious roots have a wider root angle to explore the surrounding soil surface. Additionally, some roots escape above the water surface, for example, maize and sorghum (Abiko et al. 2012; Khanthavong et al. 2021; Zhang et al. 2019a). Furthermore, in response to waterlogging, rice plants adapt by adjusting root thickness and reducing branching. These modifications create a larger channel for oxygen diffusion (due to increased thickness), minimize oxygen loss during the transport (due to shorter distance), and optimize oxygen consumption (by limiting branching).

Therefore, waterlogging conditions limit the whole root system. For example, shallow root distribution is observed in subsessile species, leading to a decrease in total root biomass (Yamauchi et al. 2019). Better root elongation to both shallow and deep soil layers contributes to increasing root biomass, and these phenotypic characteristics indicate tolerance genotypes (Khanthavong et al. 2021; Zhang et al. 2019a).

11.4 Root Response to Deficit Soil Water Stress

Deficit soil water stress is mainly caused by low rainfall or prolonged dry periods, and soil water loss is due to evaporation and transpiration related to high-temperature that decreases soil moisture. Among various environmental stresses, drought is also one of the serious stresses which has a significantly negative impact on plant biomass and yield in arid, semi-arid, and tropical regions. The effect of drought on yield depends on growth stage. Reduction in rice grain yield in intermittent and terminal drought tends to decline filled grain percentage, harvest index, and panicle number (Xangsayasane et al. 2014). Phenotyping root is one of the drought management tools as roots are more prone to drought conditions and play a significant role in the plant's survival by extracting resources from deeper soil layers to carry on several metabolic functions in the plant's body, and its phenotyping helps to understand different root traits (Wasaya et al. 2018; Wasson et al. 2012). For example, elongation rate and number of lateral roots can be decreased by high soil water content or soil density, and this can also be associated with shoot growth reduction (Bengough et al. 2011).

Plant's ability to extract water from deeper soil layers has great relevance in balancing water relation as well as carbon assimilation. The plant root distribution through the different soil layers indicates root water uptake from the soil for sustainable growth development and productivity (Coelho and Or 1999). It is important in regulating soil water use and thereby improving endurance of plants to seasonal droughts for sustainable agricultural productivity (Yu et al. 2007). Under drought, root distribution associated with deep rooting to deeper soil layers has an advantage in water uptake from further underground (Gregory 2008; Hund et al. 2009; Lynch and Wojciechowski 2015; Uga et al. 2013). Similar findings are found in maize, sorghum, millet, and rice grown under mid-dry soil conditions that these cereal crops have deep root length density associated with root plasticity (Fig. 11.1). Deep rooting serves multiple functions beyond water uptake. It is not only enhance the plant's ability to acquire water but also provides access to nutrients in sub-soil

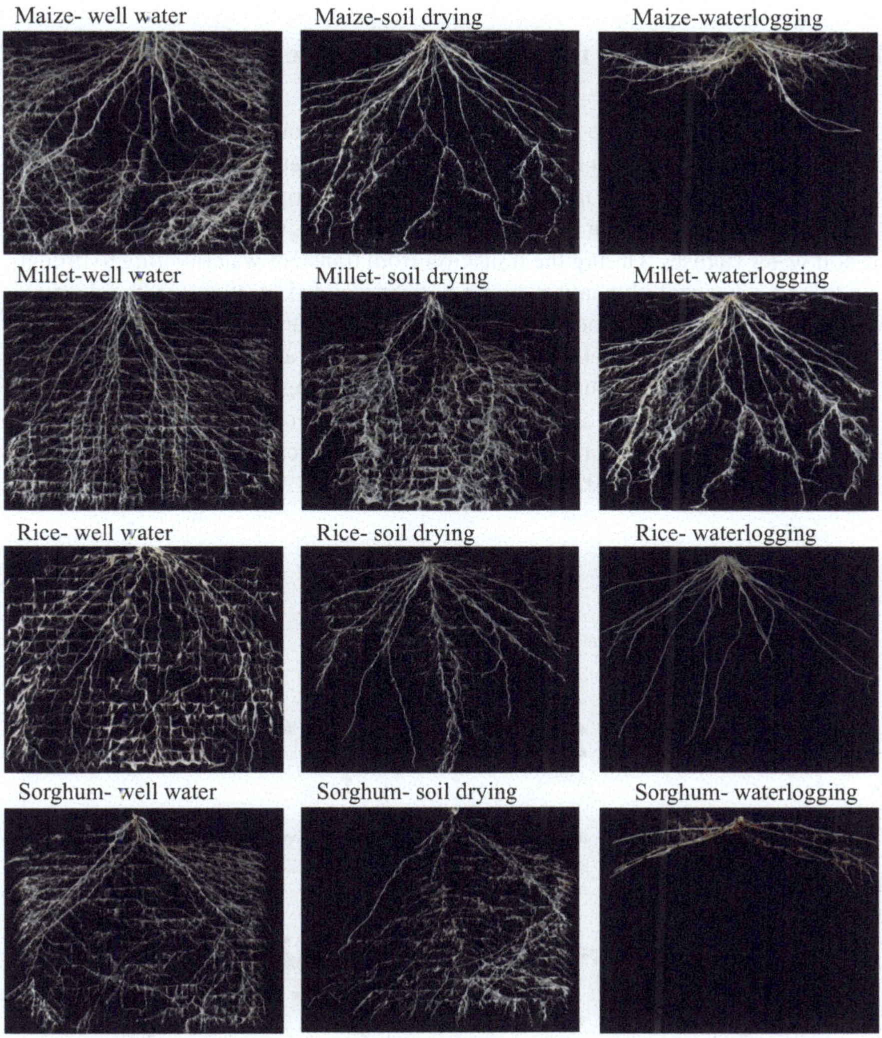

Fig. 11.1 Figure capture of cereal crops (maize, sorghum, millet, and rice) grown in a root box shows the root response to changing soil moisture contents from soil drying—well water—waterlogging. In the first column (left), under optimum soil moisture content, all crops have high root density and elongate to deeper soil layers for better accessory of nutrients. In the middle column, under limiting soil moisture content, all crops have deeper root elongation and branch roots to access the water at standing water at the deeper soil layers. The last column (right edge), root elongation of maize and sorghum have shallow expansion near soil surface to access the shallow available oxygen content for their survival. Whereas rice and millet (Banyard-millet) have better root elongation to deeper soil layers than maize and sorghum, indicating adaptation strategy to waterlogging

layers, particularly nitrogen. Thereby, the ability of crop plants such as high root density distribution to deep sub-soil layers and root deep rooting indicates traits for crop tolerant under drought conditions.

11.5 Root Response to Changing Soil Water Status

During development of crop plants, they frequently encounter fluctuating soil water conditions, transitioning from aerobic to anaerobic states throughout the growth season. Understanding the morphological and physiological modification of crop plants is a crucial discussion point. As mentioned above, roots are highly influenced by soil water statuses. During the transition from transient waterlogging to drought conditions, deep root elongation, along with associated plasticity in the deep soil layer, signifies important adaptation traits in rice plants (Suralta 2010), and it depends on variety (Kameoka et al. 2015). In contrast among the crop genotypes, the variation of root distribution responding to changing soil water status depends on crop genotypes (Fig. 11.2), and the ability of deep root distribution associated with plasticity indicates tolerant genotypes under both aerobic and anaerobic conditions (Khanthavong et al. 2021).

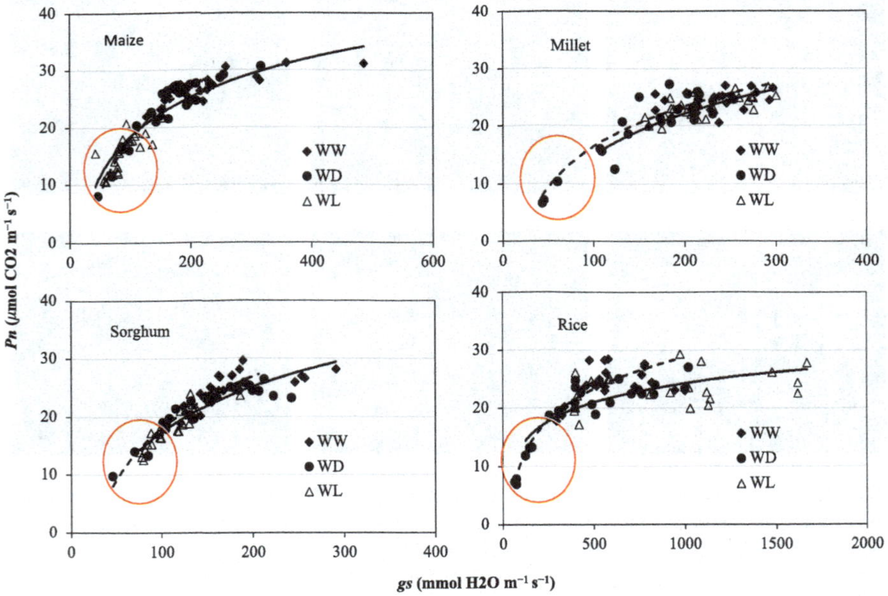

Fig. 11.2 The correlation between stomatal conductance (*gs*) and photosynthesis (*Pn*) of maize, millet, sorghum, and rice are grown under well water (WW), water deficit (WD), and waterlogging (WL). Dashed and full lines indicate low (24/15 °C) and high (34/25 °C) temperatures. Red circles indicate the impact of soil water status on *gs* related to *Pn* (adopted from Khanthavong et al. 2022)

11.6 Leaves Response to Waterlogging and Drought

11.6.1 Waterlogging

Waterlogging conditions induce deprivation of plant root respiration, which is related to insufficient energy production to maintain plant's metabolism (Loreti and Perata 2020). In addition, waterlogging also reduces root hydraulic conductance, leading to inhibition of leaf and stem growth (Else et al. 1995). Moreover, reducing root hydraulic conductance directly affects the nutrient uptake (Else et al. 1995; Smethurst et al. 2005), especially nitrogen (N) uptake (Akhtar and Nazir 2013; Azhar et al. 2022; Kaur et al. 2018; Ren et al. 2017). N content in the leaf is important to determine chlorophyll content (Herzog et al. 2016), whereas chlorophyll functions for light harvesting to improve photosynthetic system (Parkhill et al. 2001). Many sensitive plant species decrease leaf greenness (chlorophyll content) and chlorosis under waterlogging (Boru et al. 2001; Pang et al. 2004; Wei et al. 2013; Yordanova and Popova 2007). The screening of maize (*Z. mays cv.* × *Z. nicaraguensis*) tolerant to waterlogging indicates that leaf chlorosis is an important trait to consider as tolerant waterlogging (Campbell et al. 2015; Mano et al. 2006; Mano and Omori 2013). Furthermore, the degree of leaf chlorosis is not only influenced by waterlogging but also by the combination of waterlogging and light intensity, which increases severity of leaf chlorosis, especially in the lower leaf position (Goto et al. 2022). Therefore, waterlogging consequently affects the whole plant growth and yield (Jitsuyama 2017; Khanthavong et al. 2022; Promkhambut et al. 2011; Ren et al. 2014). Another characteristic of impact of waterlogging is leaf expansion, which can be determined by leaf area (Khanthavong et al. 2021; Promkhambut et al. 2010). Almost all sensitive crop plants suspend leaf growth when dealing with limiting resources supplied by roots. Limiting of leaf explanation has a negative effect on whole canopy photosynthesis and shoot biomass production, because leaf area optimization is crucial for balancing energy capture (photosynthesis) with resource allocation (biomass production) in plants.

11.6.2 Drought

The obvious symptoms of water deficit during the vegetative period are increased plant height, leaf wilting, and changes in the number and area of leaves. The leaf is an important organ for assimilation and transpiration. The plant leaf response to deficit soil water content surrounding root-zone shows a reduction in leaf area and an increase in leaf thickness and leaf angle (Manickavelu et al. 2006; Werner et al. 1999). The main reason for decreasing plant leaf area is leaf turgor pressure caused by water deficit and canopy temperature (Taiz et al. 2015). For example, rice (Patmi and Pitoyo 2020) and wheat (Werner et al. 1999) show a decrease in leaf area under water deficit. However, the leaf response to drought stress depends on crop genotypes. For example, sugarcane has marginal leaf elongation under drought stress (Misra et al. 2020), whereas wheat shows flag leaf rolling under drought stress (Willick et al. 2018). The function of leaf rolling can avoid direct sunlight to maintain water retention (Picotte et al. 2009). Leaf rolling in rice is associated with stomatal closure leading to decreased transpiration (Cal et al. 2019). Sorghum is well known for drought tolerant. Under drought conditions, sorghum has small leaves and high epicuticular wax deposition on the leaf surface, which increases leaf reflectance in the visible and near-infrared radiation and contributes to decreased transpiration (Surwenshi et al. 2010).

11.7 Relationship Between Root and Shoot Response Changing Soil Water Status: Gas Exchange and Biomass Production

The modification of root and shoot in response to water stress not only prevents plant damage, but also improves photosynthesis, growth development, and yield. In direct root measurement, functional root traits contributing to shoot growth in crop plants induce complexity (Wasson et al. 2012). The impact of abiotic stress is observed as a higher reduction of root biomass than shoot biomass (Malik et al. 2001). Thus, root to shoot ratio (*r*/*s*) is an indicator to evaluate adaptability in crop plants. For example, higher *r*/*s* of rice grown under drought conditions promotes soluble sugar in the root and higher proportion of dry matter (Xu et al. 2015); under severe drought stress, *r*/*s* of wheat (Introgression line; IL20) enhances water potential in leaves (Bacher et al. 2022), whereas *r*/*s* of chickpea is evaluated for adaptive ability to salinity (Singla and Garg 2005).

The evidence above shows how the root contributes to shoot growth under abiotic stress, especially changing in soil water status. Another essential function of the root is to activate hormone in transducing stress signals, and the main player among them is abscisic acid (ABA). ABA is well known for its functional activation of stomatal closure under flooding and drought. Drought conditions elevate ABA synthesis in the root to increase root elongation. At the same time, it is directly transported to leaf to activate stomatal closure and inhibited cell deviation, leading to inhibit shoot growth (Barnabás et al. 2008). However, under waterlogging, ABA is not directly transported from root to shoot. Instead, ethylene acts as elevating agent for ABA accumulation in the leaves under waterlogging, thereby inducing stomatal closure (Zhao et al. 2021). The dynamic of stomatal movement in response to water stress is to optimize the trade-off between assimilation and to maintain plant water status (transpiration rate) (Becklin et al. 2021). A close positive correlation among stomatal conductance, photosynthesis, and plant growth have been found under the control environments and field experiments (Fischer et al. 1998; Orzechowska et al. 2021; Wong et al. 1979). The plant's mechanism of response to water stress includes conservative stomatal closure, where the plants close their stomata faster (flooding), and non-conservative stomatal closure, where the plants close their stomata slower under drought conditions (Caine et al. 2019). Morphological metafiction roots such as deep root distribution in response to drought and waterlogging indicates tolerant contrasted among species and genotypes. Deep root distribution enhances functional water uptake ability under drought, and under waterlogging, it is associated with highly internal oxygen concentration at root tip, affecting respiration. These phenomena are significant for reactive stomata (Khanthavong et al. 2022; Suralta 2010). Conversely, sorghum (cv. SP1) with shallow root distribution has higher number of adventitious roots after 20 days of imposed waterlogging, leading to maintained stomatal opening and photosynthesis under waterlogging (Promkhambut et al. 2010). This indicates that it has a high ability of internal transport to the tip through aerenchyma formation, which enhances water uptake associated with nutrient.

Adventitious root is an essential trait which contributes to maintaining stomatal opening, photosynthesis, and biomass production under waterlogging. Under water deficit, the leaf gradually decreases in water potential with depletion of soil moisture content (Rodríguez-Gamir et al. 2019). The active stomatal aperture under drought is different from that under waterlogging. Conversely, leaf water potential is not remarkedly different under soil waterlogging (Dreyer et al. 1991). Stomatal closure during water deficit at the root-zone is an essential response to prevent water loss through leaf transpiration. During severe drought, stomatal closure causes a reduction in shoot biomass. In natural agroecosystem, drought occurring gradually depletes soil water content. The ability of plants maintaining active stomatal conductance is a mitigative mechanism of crop plants to prevent loss during drought period (Khanthavong et al. 2022). Due to current global climate change, crop plants often experience a combination between water stress and temperature variation. The optimum temperature suitable for photosynthesis is different between C_3 and C_4 species (Crafts-Brandner and Salvucci 2002; Osborne et al. 2008; Zhang et al. 2019b). Many studies have reported the effect of combinations of water stress and temperature variability on the growth and productivity of crops (Bin et al. 2010; Craufurd and Peacock 1993; Fahad et al. 2017; Heyne and Brunson 1940; Rizhsky et al. 2004). However, knowledge on the effect of various soil water status and temperature variabilities such as soil waterlogging, dry conditions, and their interactions with low and high temperatures on stomatal response among crop genotypes is scant. Furthermore, Khanthavong et al. (2022) demonstrated the effect of a combination between water stress (anaerobic to aerobic conditions) and variation of temperature by using C_3 (rice) and C_4 (maize, millet, and sorghum). The results showed that stomatal conductance, photosynthetic rate, and transpiration rate of maize and sorghum decreased in waterlogged soils, as well as aboveground biomass, regardless of temperature change, suggesting that hypoxia in the soil limited their growth. On the other hand, stomatal conductance and photosynthetic rate decreased in the dry soil at lower temperatures in maize sorghum and millet compared to rice (Fig. 11.3). This hypothesizes was related to the characteristic of the photosynthetic rate of C_4 and C_3 species, which is slow at low temperatures.

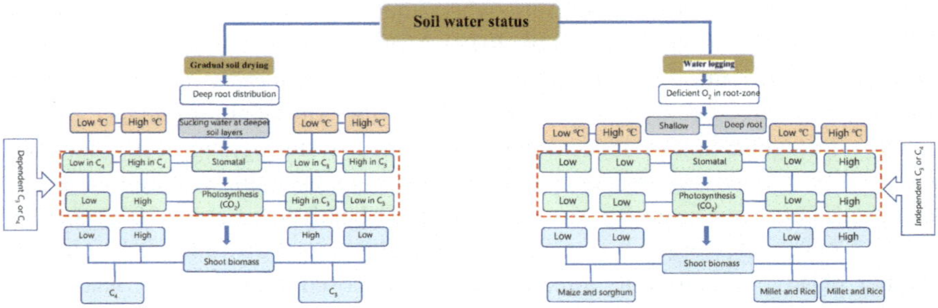

Fig. 11.3 Diagram of maize, millet, sorghum (C_4), and rice (C_3) respond to changing soil water status (adopted from Khanthavong et al. 2021, 2022). Maize and sorghum under waterlogging conditions reduce shoot biomass, presumably due to the decreased stomatal conductance and photosynthesis, which is temperature independent. In contrast, the reduction of rice and millet is also due to decreased stomatal conductance, which is temperature dependent. All crops indicate temperature-dependent stomatal conductance (at high temperature), where the *gs* of rice is lowest under high temperature. Thus, results suggest that an ability to sustain *gs* is essential for photo assimilation and maintaining leaf temperature through evapotranspiration for biomass production, a mechanism of crop avoidance to combine variable soil water status and temperature

11.8 Conclusion

Changing in soil water status directly affects plant biomass and yield in waterlogging conditions, root distribution is associated with shallow root, indicating as survival strategy, while deep root distribution represents a tolerant strategy in cereal crops. Adventitious root development under waterlogging is an essential trait that enhances internal oxygen transport to root tip for maintaining root respiration, leading to leaf photosynthesis of the plant. Deep root distribution to deeper soil layers associated with root lateral root plasticity is beneficial for available water uptake at deep soil layers. Thus, it enhances stomatal aperture, photosynthesis, and leaf cool-down through transpiration. The impact of waterlogging and drought is higher in root than shoot; therefore, crop selection for tolerant water stress by evaluating *r*/*s* are widely used. Due to current climate change, crop plants grown in the field often suffer not only water stress but also from a combination of water stress and temperature variability. Stomatal conductance measurement data correlate with photosynthesis and root distribution of cereal grown under aerobic to anaerobic condition. It suggests that maintaining ability of stomatal conductance during the water stress can be determined as tolerant strategy in crop plants.

References

Abiko T, Kotula L, Shiono K, Malik AI, Colmer TD, Nakazono M (2012) Enhanced formation of aerenchyma and induction of a barrier to radial oxygen loss in adventitious roots of *Zea nicaraguensis* contribute to its waterlogging tolerance as compared with maize (*Zea mays* ssp. mays). Plant Cell Environ 35:1618–1630

Akhtar I, Nazir N (2013) Effect of waterlogging and drought stress in plants. Int J Water Resour Environ Sci 2:34–40

Armstrong W (1979) Aeration in higher plants. In: Advances in botanical research, vol 7. Elsevier, pp 225–332

Azhar A, Asano K, Sugiura D, Kano-Nakata M, Ehara H (2022) Waterlogged conditions influence the nitrogen, phosphorus, potassium, and sugar distribution in sago palm (*Metroxylon sagu* Rottb.) at seedling stages. Plant Theory 11:710

Bacher H, Sharaby Y, Walia H, Peleg Z (2022) Modifying root-to-shoot ratio improves root water influxes in wheat under drought stress. J Exp Bot 73:1643–1654

Barnabás B, Jäger K, Fehér A (2008) The effect of drought and heat stress on reproductive processes in cereals. Plant Cell Environ 31:11–38

Becklin KM, Ward JK, Way DA (2021) Photosynthesis, respiration, and climate change. Springer

Bengough AG, McKenzie B, Hallett P, Valentine T (2011) Root elongation, water stress, and mechanical impedance: a review of limiting stresses and beneficial root tip traits. J Exp Bot 62:59–68

Berg A, de Noblet-Ducoudré N, Sultan B, Lengaigne M, Guimberteau M (2013) Projections of climate change impacts on potential C4 crop productivity over tropical regions. Agric For Meteorol 170:89–102

Bin T, Xu S-Z, Zou X-L, Zheng Y-L, Qiu F-Z (2010) Changes of antioxidative enzymes and lipid peroxidation in leaves and roots of waterlogging-tolerant and waterlogging-sensitive maize genotypes at seedling stage. Agric Sci China 9:651–661

Boru G, Van Ginkel M, Kronstad W, Boersma L (2001) Expression and inheritance of tolerance to waterlogging stress in wheat. Euphytica 117:91–98

Caine RS, Yin X, Sloan J, Harrison EL, Mohammed U, Fulton T, Biswal AK, Dionora J, Chater CC, Coe RA (2019) Rice with reduced stomatal density conserves water and has improved drought tolerance under future climate conditions. New Phytol 221:371–384

Cal AJ, Sanciangco M, Rebolledo MC, Luquet D, Torres RO, McNally KL, Henry A (2019) Leaf morphology, rather than plant water status, underlies genetic variation of rice leaf rolling under drought. Plant Cell Environ 42:1532–1544

Campbell MT, Proctor CA, Dou Y, Schmitz AJ, Phansak P, Kruger GR, Zhang C, Walia H (2015) Genetic and molecular characterization of submergence response identifies Subtol6 as a major submergence tolerance locus in maize. PLoS One 10:e0120385

Coelho EF, Or D (1999) Root distribution and water uptake patterns of corn under surface and subsurface drip irrigation. Plant Soil 206:123–136

Colmer T (2003) Long-distance transport of gases in plants: a perspective on internal aeration and radial oxygen loss from roots. Plant Cell Environ 26:17–36

Colmer T, Cox M, Voesenek L (2006) Root aeration in rice (*Oryza sativa*): evaluation of oxygen, carbon dioxide, and ethylene as possible regulators of root acclimatizations. New Phytol 170:767–778

Colmer TD, Winkel A, Kotula L, Armstrong W, Revsbech NP, Pedersen O (2020) Root O_2 consumption, CO_2 production and tissue concentration profiles in chickpea, as influenced by environmental hypoxia. New Phytol 226:373–384

Comas L, Becker S, Cruz VMV, Byrne PF, Dierig DA (2013) Root traits contributing to plant productivity under drought. Front Plant Sci 4:442

Crafts-Brandner SJ, Salvucci ME (2002) Sensitivity of photosynthesis in a C4 plant, maize, to heat stress. Plant Physiol 129:1773–1780

Craufurd P, Peacock J (1993) Effect of heat and drought stress on sorghum (Sorghum bicolor). II. Grain yield. Exp Agric 29:77–86
Den Herder G, Van Isterdael G, Beeckman T, De Smet I (2010) The roots of a new green revolution. Trends Plant Sci 15:600–607
Dreyer E, Colin-Belgrand M, Biron P (1991) Photosynthesis and shoot water status of seedlings from different oak species submitted to waterlogging. In: Annales des sciences forestières, vol 48. EDP Sciences, pp 205–214
Else MA, Davies WJ, Malone M, Jackson MB (1995) A negative hydraulic message from oxygen-deficient roots of tomato plants? (Influence of soil flooding on leaf water potential, leaf expansion, and synchrony between stomatal conductance and root hydraulic conductivity). Plant Physiol 109:1017–1024
Fahad S, Bajwa AA, Nazir U, Anjum SA, Farooq A, Zohaib A, Sadia S, Nasim W, Adkins S, Saud S (2017) Crop production under drought and heat stress: plant responses and management options. Front Plant Sci 8:1147
Fischer R, Rees D, Sayre K, Lu ZM, Condon A, Saavedra AL (1998) Wheat yield progress associated with higher stomatal conductance and photosynthetic rate, and cooler canopies. Crop Sci 38:1467–1475
Fry EL, Evans AL, Sturrock CJ, Bullock JM, Bardgett RD (2018) Root architecture governs plasticity in response to drought. Plant Soil 433:189–200
Giorgi F, Raffaele F, Coppola E (2019) The response of precipitation characteristics to global warming from climate projections. Earth Syst Dyn 10:73–89
Goto K, Yabuta S, Tamaru S, Ssenyonga P, Emanuel B, Katsuhama N, Sakagami J-I (2022) Root hypoxia causes oxidative damage on photosynthetic apparatus and interacts with light stress to trigger abscission of lower position leaves in capsicum. Sci Hortic 305:111337
Gregory PJ (2008) Plant roots: growth, activity and interactions with the soil. Wiley
Grossman JD, Rice KJ (2012) Evolution of root plasticity responses to variation in soil nutrient distribution and concentration. Evol Appl 5:850–857
Herzog M, Striker GG, Colmer TD, Pedersen O (2016) Mechanisms of waterlogging tolerance in wheat—a review of root and shoot physiology. Plant Cell Environ 39:1068–1086
Heyne E, Brunson AM (1940) Genetic studies of heat and drought tolerance in maize. J Am Soc Agron 32:803–814
Hund A, Ruta N, Liedgens M (2009) Rooting depth and water use efficiency of tropical maize inbred lines, differing in drought tolerance. Plant Soil 318:311–325
IPCC (2014) Climate change 2014: synthesis report. Geneva, IPCC
Jitsuyama Y (2017) Hypoxia-responsive root hydraulic conductivity influences soybean cultivar-specific waterlogging tolerance. Am J Plant Sci 8:770
Kameoka E, Suralta RR, Mitsuya S, Yamauchi A (2015) Matching the expression of root plasticity with soil moisture availability maximizes production of rice plants grown in an experimental sloping bed having soil moisture gradients. Plant Prod Sci 18:267–276
Kaur G, Nelson KA, Motavalli PP (2018) Early-season soil waterlogging and N fertilizer sources impacts on corn N uptake and apparent N recovery efficiency. Agronomy 8:102
Khanthavong P, Yabuta S, Asai H, Hossain M, Akagi I, Sakagami J-I (2021) Root response to soil water status via interaction of crop genotype and environment. Agronomy 11:708
Khanthavong P, Yabuta S, Malik AI, Hossain MA, Akagi I, Sakagami J-I (2022) Combinational variation temperature and soil water response of stomata and biomass production in maize, millet, sorghum and rice. Plants 11:1039
Lorbiecke R, Sauter M (1999) Adventitious root growth and cell-cycle induction in deepwater rice. Plant Physiol 119:21–30
Loreti E, Perata P (2020) The many facets of hypoxia in plants. Plant Theory 9:745
Lynch J (1995) Root architecture and plant productivity. Plant Physiol 109:7–13
Lynch JP (2007) Roots of the second green revolution. Aust J Bot 55:493–512
Lynch JP, Wojciechowski T (2015) Opportunities and challenges in the subsoil: pathways to deeper rooted crops. J Exp Bot 66:2199–2210

Malik AI, Colmer TD, Lambers H, Schortemeyer M (2001) Changes in physiological and morphological traits of roots and shoots of wheat in response to different depths of waterlogging. Funct Plant Biol 28:1121–1131

Manickavelu A, Nadarajan N, Ganesh S, Gnanamalar R, Chandra Babu R (2006) Drought tolerance in rice: morphological and molecular genetic consideration. Plant Growth Regul 50:121–138

Mano Y, Omori F (2013) Flooding tolerance in interspecific introgression lines containing chromosome segments from teosinte (*Zea nicaraguensis*) in maize (*Zea mays* subsp. mays). Ann Bot 112:1125–1139

Mano Y, Omori F, Muraki M, Takamizo T (2005) QTL mapping of adventitious root formation under flooding conditions in tropical maize (*Zea mays* L.) seedlings. Breed Sci 55:343–347

Mano Y, Muraki M, Takamizo T (2006) Identification of QTL controlling flooding tolerance in reducing soil conditions in maize (*Zea mays* L.) seedlings. Plant Prod Sci 9:176–181

McMichael B, Quisenberry J (1993) The impact of the soil environment on the growth of root systems. Environ Exp Bot 33:53–61

Misra V, Solomon S, Mall A, Prajapati C, Hashem A, Abd Allah EF, Ansari MI (2020) Morphological assessment of water stressed sugarcane: a comparison of waterlogged and drought affected crop. Saudi J Biol Sci 27:1228–1236

Negi S, Sukumar P, Liu X, Cohen JD, Muday GK (2010) Genetic dissection of the role of ethylene in regulating auxin-dependent lateral and adventitious root formation in tomato. Plant J 61:3–15

Nguyen T-N, Tuan PA, Mukherjee S, Son S, Ayele BT (2018) Hormonal regulation in adventitious roots and during their emergence under waterlogged conditions in wheat. J Exp Bot 69:4065–4082

Orzechowska A, Trtílek M, Tokarz KM, Szymańska R, Niewiadomska E, Rozpądek P, Wątor K (2021) Thermal analysis of stomatal response under salinity and high light. Int J Mol Sci 22:4663

Osborne CP, Wythe EJ, Ibrahim DG, Gilbert ME, Ripley BS (2008) Low temperature effects on leaf physiology and survivorship in the C3 and C4 subspecies of *Alloteropsis semialata*. J Exp Bot 59:1743–1754

Pang J, Zhou M, Mendham N, Shabala S (2004) Growth and physiological responses of six barley genotypes to waterlogging and subsequent recovery. Aust J Agric Res 55:895–906

Pareek A, Sopory SK, Bohnert H (2009) Abiotic stress adaptation in plants. Springer

Pareek A, Dhankher OP, Foyer CH (2020) Mitigating the impact of climate change on plant productivity and ecosystem sustainability, vol 71. Oxford University Press, p 451

Parkhill JP, Maillet G, Cullen JJ (2001) Fluorescence-based maximal quantum yield for PSII as a diagnostic of nutrient stress. J Phycol 37:517–529

Patmi Y, Pitoyo A (2020) Effect of drought stress on morphological, anatomical, and physiological characteristics of Cempo Ireng cultivar mutant rice (*Oryza sativa* L.) strain 51 irradiated by gamma-ray. J Phys Conf Ser 1436:012015

Pedersen O, Nakayama Y, Yasue H, Kurokawa Y, Takahashi H, Heidi Floytrup A, Omori F, Mano Y, David Colmer T, Nakazono M (2021) Lateral roots, in addition to adventitious roots, form a barrier to radial oxygen loss in *Zea nicaraguensis* and a chromosome segment introgression line in maize. New Phytol 229:94–105

Picotte JJ, Rhode JM, Cruzan MB (2009) Leaf morphological responses to variation in water availability for plants in the *Piriqueta caroliniana* complex. Plant Ecol 200:267–275

Promkhambut A, Younger A, Polthanee A, Akkasaeng C (2010) Morphological and physiological responses of sorghum (*Sorghum bicolor* L. Moench) to waterlogging. Asian J Plant Sci 9:183

Promkhambut A, Polthanee A, Akkasaeng C, Younger A (2011) Growth, yield and aerenchyma formation of sweet and multipurpose sorghum (*Sorghum bicolor* L. Moench) as affected by flooding at different growth stages. Aust J Crop Sci 5:954–965

Raskin I, Kende H (1984) Regulation of growth in stem sections of deep-water rice. Planta 160:66–72

Ren B, Zhang J, Li X, Fan X, Dong S, Liu P, Zhao B (2014) Effects of waterlogging on the yield and growth of summer maize under field conditions. Can J Plant Sci 94:23–31
Ren B, Dong S, Zhao B, Liu P, Zhang J (2017) Responses of nitrogen metabolism, uptake and translocation of maize to waterlogging at different growth stages. Front Plant Sci 8:1216
Rizhsky L, Liang H, Shuman J, Shulaev V, Davletova S, Mittler R (2004) When defense pathways collide. The response of Arabidopsis to a combination of drought and heat stress. Plant Physiol 134:1683–1696
Rockström J, Karlberg L, Wani SP, Barron J, Hatibu N, Oweis T, Bruggeman A, Farahani J, Qiang Z (2010) Managing water in rainfed agriculture—the need for a paradigm shift. Agric Water Manag 97:543–550
Rodríguez-Gamir J, Xue J, Clearwater MJ, Meason DF, Clinton PW, Domec JC (2019) Aquaporin regulation in roots controls plant hydraulic conductance, stomatal conductance, and leaf water potential in *Pinus radiata* under water stress. Plant Cell Environ 42:717–729
Sakagami J-I, Kawano N (2011) Survival of submerged rice in a flood-prone region of West Africa. Tropics 20:55–66
Sasidharan R, Bailey-Serres J, Ashikari M, Atwell BJ, Colmer TD, Fagerstedt K, Fukao T, Geigenberger P, Hebelstrup KH, Hill RD (2017) Community recommendations on terminology and procedures used in flooding and low oxygen stress research. New Phytol 214:1403–1407
Singla R, Garg N (2005) Influence of salinity on growth and yield attributes in chickpea cultivars. Turk J Agric For 29:231–235
Smethurst CF, Garnett T, Shabala S (2005) Nutritional and chlorophyll fluorescence responses of lucerne (*Medicago sativa*) to waterlogging and subsequent recovery. Plant Soil 270:31–45
Steffens B, Rasmussen A (2016) The physiology of adventitious roots. Plant Physiol 170:603–617
Suralta RR (2010) Plastic root system development responses to drought-enhanced nitrogen uptake during progressive soil drying conditions in rice. Philipp Agric Sci 93:458–462
Suralta R, Inukai Y, Yamauchi A (2010) Shoot dry matter production in relation to root growth, oxygen transport and water uptake under transient soil moisture stresses in rice. Plant Soil 332:87–104
Surwenshi A, Chimmad V, Jalageri B, Kumar V, Ganapathi M, Nakul H (2010) Characterization of sorghum genotypes for physiological parameters and yield under receding soil moisture conditions. Res J Agric Sci 1:242–244
Taiz L, Zeiger E, Møller IM, Murphy A (2015) Plant physiology and development. Sinauer Associates Incorporated
Uga Y, Sugimoto K, Ogawa S, Rane J, Ishitani M, Hara N, Kitomi Y, Inukai Y, Ono K, Kanno N (2013) Control of root system architecture by DEEPER ROOTING 1 increases rice yield under drought conditions. Nat Genet 45:1097–1102
United Nations (2019) World population prospects 2019: highlights. Population Division of the UN Department of Economic and Social Affairs
Vadez V (2014) Root hydraulics: the forgotten side of roots in drought adaptation. Field Crop Res 165:15–24
Vidoz ML, Loreti E, Mensuali A, Alpi A, Perata P (2010) Hormonal interplay during adventitious root formation in flooded tomato plants. Plant J 63:551–562
Wasaya A, Zhang X, Fang Q, Yan Z (2018) Root phenotyping for drought tolerance: a review. Agronomy 8:241
Wasson AP, Richards R, Chatrath R, Misra S, Prasad SS, Rebetzke G, Kirkegaard J, Christopher J, Watt M (2012) Traits and selection strategies to improve root systems and water uptake in water-limited wheat crops. J Exp Bot 63:3485–3498
Weaver JE, Jean FC, Crist JW (1922) Development and activities of roots of crop plants: a study in crop ecology. Carnegie Institution of Washington
Wei W, Li D, Wang L, Ding X, Zhang Y, Gao Y, Zhang X (2013) Morpho-anatomical and physiological responses to waterlogging of sesame (*Sesamum indicum* L.). Plant Sci 208:102–111

Werner C, Correia O, Beyschlag W (1999) Two different strategies of Mediterranean macchia plants to avoid photoinhibitory damage by excessive radiation levels during summer drought. Acta Oecol 20:15–23

Willick IR, Lahlali R, Vijayan P, Muir D, Karunakaran C, Tanino KK (2018) Wheat flag leaf epicuticular wax morphology and composition in response to moderate drought stress are revealed by SEM, FTIR-ATR and synchrotron X-ray spectroscopy. Physiol Plant 162:316–332

Wong S, Cowan I, Farquhar G (1979) Stomatal conductance correlates with photosynthetic capacity. Nature 282:424–426

Xangsayasane P, Fukai S, Mitchell J, Jongdee B, Jothityangkoon D, Pantuwan G, Inthapanya P (2014) Genotypic performance in multi-location on-farm trials for evaluation of different on-station screening methods for drought-prone rainfed lowland rice in Lao PDR. Field Crop Res 160:1–11

Xu W, Cui K, Xu A, Nie L, Huang J, Peng S (2015) Drought stress condition increases root to shoot ratio via alteration of carbohydrate partitioning and enzymatic activity in rice seedlings. Acta Physiol Plant 37:1–11

Yamauchi T, Colmer TD, Pedersen O, Nakazono M (2018) Regulation of root traits for internal aeration and tolerance to soil waterlogging-flooding stress. Plant Physiol 176:1118–1130

Yamauchi T, Abe F, Tsutsumi N, Nakazono M (2019) Root cortex provides a venue for gas-space formation and is essential for plant adaptation to waterlogging. Front Plant Sci 10:259

Yamauchi T, Noshita K, Tsutsumi N (2021) Climate-smart crops: key root anatomical traits that confer flooding tolerance. Breed Sci 71:51

Yordanova RY, Popova LP (2007) Flooding-induced changes in photosynthesis and oxidative status in maize plants. Acta Physiol Plant 29:535–541

Yu G-R, Zhuang J, Nakayama K, Jin Y (2007) Root water uptake and profile soil water as affected by vertical root distribution. Plant Ecol 189:15–30

Zhang F, Zhu K, Wang Y, Zhang Z, Lu F, Yu H, Zou J (2019a) Changes in photosynthetic and chlorophyll fluorescence characteristics of sorghum under drought and waterlogging stress. Photosynthetica 57:1156–1164

Zhang R, Zhou Y, Yue Z, Chen X, Cao X, Ai X, Jiang B, Xing Y (2019b) The leaf-air temperature difference reflects the variation in water status and photosynthesis of sorghum under waterlogged conditions. PLoS One 14:e0219209

Zhao Y, Zhang W, Abou-Elwafa SF, Shabala S, Xu L (2021) Understanding a mechanistic basis of ABA involvement in plant adaptation to soil flooding: the current standing. Plant Theory 10:1982

The manufacturer's authorised representative in the EU is Springer
Nature Customer Service Centre GmbH, Europaplatz 3, 69115 Heidelberg,
Germany. If you have any concerns regarding our products, please
contact ProductSafety@springernature.com

Printed and bound by CPI Group (UK) Ltd, Croydon, CR0 4YY
01/10/2026
02225904-0002